OXIDATION OF PETROCHEMICALS: CHEMISTRY AND TECHNOLOGY

OXIDATION
OF PETROCHEMICALS:
CHEMISTRY AND TECHNOLOGY

THEODORE DUMAS, P.Eng., M.Sc.
WALTER BULANI, Ph.D.

(University of Western Ontario, Canada)

A HALSTED PRESS BOOK

JOHN WILEY & SONS
New York — Toronto

PUBLISHED IN THE U.S.A. AND CANADA BY
HALSTED PRESS
A DIVISION OF JOHN WILEY & SONS, INC., NEW YORK

Library of Congress Cataloging in Publication Data

Dumas, Theodore.
   Oxidation of petrochemicals.

   "A Halsted Press book."
   Includes bibliographical references.
   1. Oxidation.   2. Petroleum chemicals.
I. Bulani, Walter, joint author. II. Title.
TP690.45.D85   661'.804   74-11232
ISBN 0-470-22480-0

WITH 61 ILLUSTRATIONS AND 24 TABLES

© APPLIED SCIENCE PUBLISHERS LTD 1974

Printed in Great Britain by J. W. Arrowsmith Ltd, Bristol, England

To
Our parents and dear friends
and
all who seek, as we do,
the liberty of true knowledge
and
the knowledge of true liberty

Preface

At the time of this writing there is no publication which deals comprehensively with the chemistry and technology of the oxidation of petrochemicals. We believe that this alone justifies us in adding to the rapidly growing number of new books on technology. Since we have abstracted the most recent information about various process technologies from some 460 references, noted their technological, economic and chemical similarities and related these whenever possible to the corresponding fundamental processes, we feel doubly confident that this book will provide a unifying body of knowledge to every student of chemistry and chemical technology. The researchers, engineers and business executives who work in the special field of organic oxidation chemistry should find this exposition particularly helpful to their careers.

We recognise the difficulty in reporting on a rapidly developing field and trust that any omissions or other inadequacies in this work will encourage the perceptive reader to add and improve upon our efforts by bringing these to our attention.

Contents

CHAPTER 1

Scope of Petrochemical
Oxidation Processes

1.1 INTRODUCTION

The recent development of processes to produce millions of tons per year of a variety of petrochemicals owes its stimulus primarily to the tremendous growth rate and abundance of byproducts such as ethylene, propylene, benzene, toluene, xylene, naphthalene, etc. from the petroleum refining industry. The commercial successes of these processes, however, must be ascribed appropriately to those chemists and engineers who in the last few decades, developed the understanding and the technology of homogeneous and heterogeneous catalytic oxidation processes. The progress in both of these areas and fields has been particularly remarkable during the past decade.

Although progress in science and in technology are closely interrelated, it often proceeds independently and irregularly in both areas. Thus in the case of catalytic oxidation in the homogeneous liquid phase the significant improvements in commercial processes during the past 10-12 years may be ascribed primarily to the vast improvements which have been made in the understanding of ionic and ligand effects on chemical kinetics. This understanding has allowed the development of technological processes which are simple yet highly selective and capable of producing products of high purity and yield under automatic control. On the other hand the improvements in the manufacture of phthalic anhydride from naphthalene or xylene may be attributed primarily to advances in the technology and control of fluidised bed reactors.

A partial appreciation of the dynamic character of the evolution of various products (and hence related processes) can be obtained from the data given in Table 1.

The real dynamism lies however in the area which might be best termed techno-economics. The cost parameters of technological processes are such that new technology can make new processes obsolete as readily as old processes. A comparison of the overall technical and economic characteristics of different manufacturing processes indicates clearly that

1

TABLE 1
Production of some basic petrochemicals (in 10^6 tons per year)

Chemical	Estimated 1970 level		Estimated 1975 level		Reference
	World	USA	World	USA	
Ethylene	16·6	7	30·5	11	1-5
Ethylene oxide		1·35		1·8	1-5
Styrene	0·635		0·950		1-5
Propylene	9·3	4·15	15·4	5·9	6-9
Acrylonitrile	0·590		0·935		6-9
Isopropyl benzene	0·280		0·545		6-9
Benzene	8·5		13·5		9
Toluene	1·8				9
Xylene	1·7				9

oxidation in the homogeneous liquid phase occupies a prominent position on a worldwide basis. It suggests moreover that, although the hydroperoxide process for producing phenol from cumene may be expected to continue to hold first place for some time to come, the decarboxylation of benzoic acids produced by homogeneous liquid phase oxidations can gain dominance. On the other hand it appears even more certain that liquid phase oxidation processes to produce acrolein or acrylic acid by the direct ammoxidation of propylene will not be competitive within five years.

A further appreciation and perspective of the growing importance of innovative oxidation processes in the petrochemical industry can be attained from a brief overview of the topics treated in this book.

1.2   CATALYTIC OXIDATION IN THE HOMOGENEOUS LIQUID PHASE

It has already been stated that the commercial success of homogeneous liquid phase oxidation processes can be attributed to the remarkable progress which has been made in the past decade in elucidating the mechanisms of homogeneous liquid phase catalytic reactions. Thus in this chapter we show that the ligand type coordination compounds which the transition elements form with olefins, alcohols, ketones and acids, control and catalyse the oxidation reactions. Similarly we also show that metallic ions combine with acetates, propionates, butyrates, naphthenates, benzoates, etc. to catalyse oxidation reactions by the homolytic (radical) mechanism.

Among the processes based on the catalytic system $PbCl_2$-$CuCl_2$ we discuss the manufacture of vinyl acetate and acetaldehyde from ethylene, acetone from propylene and methyl ethyl ketone from n-butene. Under metal ion catalytic processes we treat in detail the manufacture of acetic acid and acetic anhydride from acetaldehyde, benzoic acid from toluene, phenol from benzoic acid, cyclohexanol from cyclohexane, adipic acid from the latter two chemicals and terephthalic acid and its anhydride from p- and o-xylenes respectively.

In each case we have attempted to collate full data on the reaction temperature, pressure and time, the type and nature of catalyst and other technological aspects which are significant to the optimisation of the process. The technoeconomic comparative study of the manufacture of phthalic anhydride however has been deferred to the chapter on heterogeneous catalytic processes. This practice is repeated whenever the product is manufactured predominantly by a more competitive process.

We conclude this chapter by pointing out that some oxidation processes function well in the homogeneous liquid phase in the absence of catalysts provided certain starters are present. The oxidation of isopropanol to acetone and of propylene and n-butylene to acetic acid are examples. The special role of hydroperoxide chemistry in the manufacture of phenol and acetone from cumene and of styrene and propylene oxide from ethyl benzene is treated, also.

1.3 HETEROGENEOUS CATALYTIC OXIDATION OF AROMATICS, ALKYL BENZENES AND OLEFINS

In this chapter we review the details of fixed and fluid phase processes for the manufacture of phthalic anhydride and assess the advantages of each process and compare these with one based on oxidation in the homogeneous liquid phase.

The manufacture of maleic anhydride by partial oxidation of benzene appears to be more important than by partial oxidation of C_4 fractions.[10-14] Since this is not dealt with later the following brief details are given now for completeness. Scientific Design Co. Inc. and Rührohl-Farbenfabriken Bayer A.G. have processes based on the former.[14] The theoretical yield from 50·8 kg of benzene is 56·9 kg of maleic anhydride. The yields obtained in practice range between 67 and 72% of this. The reaction is carried out at temperatures ranging between 425 and 450°C at pressures slightly above one atmosphere in tubular reactors. The catalyst consists of mixed oxides on a support (about 2·5% MoO_3, 5·2% P_2O_5 and 0·6% P_2O_5). The volumetric throughput of the reactant mixture ($\sim$1·2 mole percent benzene in air) is of the order of 2500 volumes per volume of catalyst per hour. Further particulars on the kinetics of this reaction can be found in references 15 to 19. Particulars on catalysts are given in references 20-26. Both alumina and silica supports are used as well as certain promoting elements.

The production of maleic anhydride by partial oxidation of C_4 fractions also requires the presence of a great excess of air (volumetric ratio of about 75:1) to maintain the reaction temperature between 400 and 475°C. The C_4 fractions utilised generally contain about 80% of 1- and 2-butylene. As an example butylene at a concentration of about 1% by volume when passed through a catalyst at 410°C at a space velocity of 3200 hours^{-1} was found to produce 38 grams of maleic anhydride per litre of catalyst per hour with an approximate yield of 55%. The productivity and yield are highly dependent upon the type and nature of catalyst. A catalyst based on Mo-V-P in the ratio of 9:3:1 was found to produce 0·4 kg of anhydride per kilogram of catalyst[27] while mixed catalyst based on oxides of Mo-Co, V-P or Mo-U with Ti and additions of boric or phosphoric acid are reported to give high conversions (93%) and good yields.[28] Catalysts based on V_2O_5 and H_3PO_4 give conversions of about 97% with yields of about 45%[29] while those based on Li_3PO_4 give about the same conversion (96%) with significantly higher yield ($\sim$58%).[30] Mitsubishi Chemical Industries Ltd operates plants based on this process.[31]

In our review of the direct oxidation of olefins we discuss the special selective character of silver based catalysts for converting ethylene to its oxide and compare its (scientific design) commercial process with those based on ethylene chlorohydrin (Shell Development Co., Nipon Shokubai Kagaku Kogio Co. Ltd and Chemische Werke Hullo A.C.). Acrylic acid is another large volume product which can be obtained by direct oxidation of an olefin (propylene). Here we have clarified the basis of the three types of commercial processes and demonstrate a superiority for the direct process. The techno-economic comparison of the two-step process where acrolein is separated out first and then oxidised to acrylic acid with the two-phase process where the acrolein is not separated out but is further oxidised in a second reactor by the direct oxidation process is instructive on how effective selective control of reactions can be achieved through appropriate manipulation of the temperature, pressure and catalyst contact time of the reactants.

1.4 CATALYTIC DEHYDROGENATION OF ALKYL BENZENES, ALKANES AND ALKENES

The manufacture of styrene from ethyl benzene, divinylbenzene from diethylbenzene, methyl styrene from isopropyl benzene, isopropenyl toluene from isopropyl toluene and isopropenyl xylene from isopropyl xylene are all based upon oxidative dehydrogenation chemistry. Special emphasis has been given to styrene because of its high production level (~ 4 million tons per year).

The influences of water vapour, benzene and styrene content of the reactant mixture, of catalyst composition, structure, and state of regeneration and of temperature on the conversion of ethyl benzene to styrene are all examined in considerable detail. The use of halogens or compounds of sulphur in lieu of oxygen to effect the dehydrogenation at lower temperatures is noted, too. The important technological parameters for the manufacture of the other products previously mentioned are given, also.

The manufacture of butadiene from butane-butylene mixtures and the manufacture of isoprene from isopentane-isopentene mixtures are sufficiently similar that there is a possibility of their simultaneous production in the same installation from the appropriate C_4 and C_5 mixture. This conclusion is supported by an examination and comparison of the effects of reactant temperature, pressure, space velocity, catalyst type and oxidant composition (oxygen, iodine, chlorine, bromine or sulphur) on conversion and yield of both products from the respective reactants. The similarities of these processes are evident under both fixed and fluid phase operations. The chief difficulty of obtaining a product of high purity arises in each case from the complex problems posed by component separation.

1.5 AMMOXIDATION OF OLEFINS

The oxidative ammonolysis or ammoxidation of aliphatic or aromatic hydrocarbons can be given by the generalised Andrussov reaction

$$R—CH_3 + 1·5O_2 + NH_3 \rightarrow RCN + 3H_2O$$

This reaction is highly exothermic and irreversible. The age old desire of chemists to introduce nitrogen into an organic molecule by treating it directly with ammonia has been realised on a grand scale in the manufacture of acrylonitrile from propylene. The technological aspects of this process are reviewed to demonstrate once again the special control and selectivity which can be achieved by appropriate manipulation of the reactant temperature, pressure and space velocity over suitable catalytic materials. The mechanical stability of the catalysts in this case is of prime importance.

Among the characteristics of the process of ammoxidation of olefins which are covered in some detail the following are equally valid when the process is applied to aromatic hydrocarbons, naphthanes or heterocyclic-alkyls.[32-34]

1 The concentration of ammonia has a strong influence over the reaction.

2 The reaction temperatures are high and special design and operation is essential to remove excess heat. The direct injection of water into the reactor has been shown to facilitate effective temperature control and to permit higher yields to be obtained by reducing the rate and degree of catalyst deactivation.

3 The catalysts employed are based generally on mixtures of V_2O_5 and Cr_2O_3 on alumina supports.

4 Although S or SO_2 can be used in lieu of oxygen to effect the reaction at lower temperatures, these oxidants raise great corrosion problems.

5 The reaction must be carried out with care to ensure that the explosion limits of mixtures of H_2 and O_2 are not exceeded. Usual concentrations are <2 vol. %.

6 It has been demonstrated that ammoxidation of lower alkanes initially involves a dehydrogenation step[35] and can be highly selective. The ammoxidation of higher alkanes is less selective and leads to increasing fractions of secondary products as the chain length increases. Both the selectivity and yield can be improved appreciably by conducting these reactions in the liquid phase.[36-38]

1.6   SUMMARY

This outline and overview of the special role and significance of specific commercial oxidation processes, like the text that follows, touches only upon those topics which in our judgement will aid the reader to understand this rapidly growing field through direct association or referral to more detailed discussions. The readers are encouraged to seek out the reference articles or other appropriate texts whenever in the interests of brevity we have failed to convey adequate clarity.

CHAPTER 2

Catalytic Oxidation in the Homogeneous Liquid Phase

2.1 CATALYST TYPES AND SPECIFICITY OF REACTION

During the past 10-12 years the progress made in the knowledge and synthesis of catalysts for oxidation in the homogeneous liquid phase has surpassed that realised in the last 50 years for catalysts used in hetero-geneous reactions. This may be attributed to the greater ease of detecting and identifying intermediate products due generally to the milder conditions of the reactions which permit simpler technology and hence greater selectivity and reproducibility.

2.1.1 Coordination Complex Catalysts

Transition metals form two types of compounds with organic substances. Those in which a bond forms between the metal and the organic chemical are generally unstable and are of minor practical importance. Examples of this type of compound are dimethyl platinum, dimethyl manganese and tetramethyl titanium. On the other hand these metals also form compounds by sharing the π electrons of the organic molecules. In this case the d-orbitals of the metal are involved in the bonding and stable co-ordination complexes of great commercial importance are formed.

The first compound of the latter category ever prepared was probably Zeise Salt which was made in 1827 in accordance with the following reaction:

$$K_2PtCl_4 + C_2H_4 \rightarrow KCl + K[PtC_2H_4Cl_3]$$

Systematic study of this type of compound however was deferred until 1951 when 'ferrocene' or dicyclo-pentadiene-iron was made and shown to have a bipyramidal sandwich structure. The evidence for the latter appears sound since the analysis of infrared and Raman spectrophoto-metric data along with that obtained by X-ray and electron diffraction is consistent with the postulate that the Fe atom is situated between two parallel cyclopentadiene molecules in such a manner that it is equidistant (1·41 Å) from each carbon atom as shown in Figure 1.

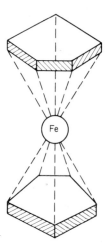

Fig. 1   Ferrocene.

The extensive studies of these complexes resulted in their rapid and wide utilisation as catalysts for large scale commercial oxidation processes in the liquid phase. The ones which show the greatest utility in this regard involve the transition elements of group VIII especially Fe, Co, Ni, Ru, Rh, Pd, Os, Ir and Pt. The following comments on the structure, stability, reactivity and selectivity of these complexes should assist the reader unfamiliar with these compounds to appreciate more fully their potential and role in regulating oxidative chemical reactions.

Generally the outer electron orbital of an atom differs from that of the element preceding it in the periodic table. However in the case of the so-called transition elements with atomic numbers 21 to 30, 39 to 48, 57 to 80 and 89 to 98 the d-orbitals of the next inner shell are at a lower energy level and are filled first. The characteristic property of these elements is consequently one of easily variable valence. Various oxidation states occur with minimal differences in energy. Hence the transition elements are particularly capable of breaking or stabilising coordinative bonds. That is, these elements interchange readily between the states metal-metal salt and metal or metal salt and coordination complex and this accounts for their great utility in homogeneous liquid phase oxidation reactions.

A typical series of coordination complexes is exemplified by the cobalt chloro-amino compounds.

1   Trichloro hexaminocobalt $[Co(NH_3)_6]Cl_3$ in which the central Co^{3+} ion, bound to six neutral NH_3 *ligands*, forms the stable complex ion.

2   Dichloro pentaminochlorocobalt $[Co(NH_3)_5Cl]Cl_2$ in which the stable ion complex consists of the central Co^{3+} 5 neutral NH_3 *ligands* and one Cl^- *ion ligand*.

3   Chloro tetraaminodichlorocobalt $[Co(NH_3)_4Cl_2]Cl$ in which case the stable complex now contains two Cl^- ions and only one chlorine per molecule is ionisable.

In each case the coordination number of the Co^{3+} ion is 6. However in cases 2 and 3 the chloride ions from the second sphere of attraction have displaced an NH_3 molecule in the first sphere of attraction. Clearly the NH_3 molecules also may be replaced by water or other molecules which are not sterically restrictive.

Both X-ray and chemical evidence indicate that complexes with a coordination number of 2 are linear, those with coordination number of 4 are square planar while those with a coordination number of 6 are octahedral.

The bonding found in these complexes is said to result from hybridisation of the dsp orbitals. Although quantum mechanical arguments may be employed to explain, support and clarify the electronic theory, the best appreciation of the nature of this bonding can be obtained by visualising the geometric requirement to bring 2, 4, 6, etc. atoms or molecules into close and identical proximity of a central ion.

The stability of such complexes may be expressed best in terms of their dissociation or association constants. For example the latter for $Ag(CN)_2^-$ is about 10^{21} whereas for the $Co(NH_3)_6^{3+}$ complex it is 10^{12} while that of $Co(NH_3)_6^{2+}$ is only 10^2.

It must be noted that ligands always possess a pair of non-participating electrons which form bonds with the central atom. This gives rise to complexes such as $[Fe(CN)_6]^{4-}$ and $[Fe(CN)_6]^{3-}$ as well as stoichiometrically identical but chemically dissimilar salts such as $[Co(NH_3)_5SO_4]^+Br^-$ and $[Co(NH_3)_5Br]^{2+}SO_4^{2-}$.

Organic compounds which contain O, N, S, P, S, As, etc. also can function as electron donors and therefore act in a manner similar to that of the inorganic molecules or ions to form ligand bonds. In this case the complexes are referred to as chelates whenever the organic component has two or more functional groups which occupy the coordinative positions of the central ion. The most common examples of such chelating-ligands are the poly acid and poly amino anions.

Olefins and aromatic compounds as alluded to at the beginning of this chapter also have the capacity to act as ligands through sharing their π electrons with the central ion. A practical example of the latter may be found in the Wacker process of oxidising ethylene to the aldehyde using $PdCl_2/CuCl_2$ as a catalyst.

Although the chemistry of this process may be described by the following sequence of reactions, the evidence is such that an intermediate ethylene palladium coordination complex seems likely.[44]

(a) $CH_2{=}CH_2 + PdCl_2 + H_2O \rightarrow CH_3CH{=}O + Pd + 2HCl$
(b) $Pd + 2CuCl_2 \rightarrow PdCl_2 + 2CuCl$
(c) $2CuCl + 2HCl + \frac{1}{2}O_2 \rightarrow 2CuCl_2 + H_2O$

The following mechanism is suggested to account for reaction (a) above:[44]

$$PdCl_4^= + C_2H_4 \rightleftharpoons [PdCl_3C_2H_4]^- + Cl^-$$

$$[PdCl_3C_2H_4]^- + H_2O \rightleftharpoons [PdCl_2(H_2O)C_2H_4] + Cl^-$$

$$[PdCl_2(H_2O)C_2H_4] + H_2O \rightarrow [PdCl_2(OH)C_2H_4]^- + H_3O^+$$

Thus if the latter mechanism is correct the oxidation of the ethylene is effected not by molecular oxygen but by the oxygen from water. The molecular oxygen is utilised to oxidise cuprous to cupric ions which in turn oxidise palladium metal to the palladium ion. The latter forms the aquo-ethylene complex which dissociates readily due to the ease of valence changes. Note that the mechanism suggests a nucleophilic attack on the double bond which is believed to be made more labile upon complexing with the Pd^{2+} ion. The hydrated alkane II permits the transfer of a hydrogen atom by reverting to structure III and the organic ion formed on dissociation of the complex is stabilised by losing a proton to form the aldehyde.

In oxidising ethylene to vinyl acetate the latter reaction would be[45]

$$CH_3\overset{+}{C}H-O-\overset{\displaystyle O}{\overset{\|}{C}}-CH_3 \rightarrow CH_2=CH-O-\overset{\displaystyle O}{\overset{\|}{C}}-CH_3 + H^+$$

The fact that Pd^{2+} is a more active catalyst than other transition metals for this reaction[46] suggests that it reduces the activation energy required for the nucleophilic attack to a greater degree.

Palladium salts are used also as catalysts in other homogeneous liquid phase oxidation reactions. The oxidative carbonylation of ethylene with carbon monoxide and oxygen to form acrylic acid and β-acetoxy propionic acid probably involves the following complexing reactions.[47]

$$PdCl_2 + 2CO + H_2O \rightarrow [Cl_2\overset{\displaystyle CO}{\overset{|}{P}}d-\overset{\displaystyle O}{\overset{\|}{C}}-OH]-C_2H_4 \rightarrow$$

$$[Cl_2\overset{\displaystyle C_2H_4}{\overset{|}{P}}d-\!\!-\!\!-\overset{\displaystyle O}{\overset{\|}{C}}-OH]^- \overset{CO}{\rightarrow} Cl_2\overset{\displaystyle CO}{\overset{|}{P}}d-C_2H_4-\overset{\displaystyle O}{\overset{\|}{C}}-OH \quad etc.$$

Although Pd ions form complexes with higher olefins the olefin reactivity is found to decrease in the order ethylene-propylene-butene-1. Acetone is formed from propylene oxidation and from butene the product is methyl ethyl ketone.

Similarly the oxidation of *derivatives* of styrene produces *derivatives* of acetophenone and of phenylacetaldehyde. The relative yield of the products produced is determined by the nature of the substituent groups.[48]

2.1.2   Catalysts of the Metallic Ion Type

Catalysts of the metal ion type are represented by diverse materials such as metals, metal oxides, metal organic acid salts and ammonium salts of metal acids. The metal salts most commonly employed are acetates and propionates of Co, Ni, Cr and Mn. Long chain organic acid as well as aromatic acid salts also are effective although generally more difficult to prepare and more costly to use.

Regardless of the nature of metal salt involved catalysts of this type regulate or control the oxidative reactions in a similar way. The mechanism which appears most consistent with all experimental data has been termed homolytic and involves three steps, free radical initiation, chain propagation and chain termination. The homolytic mechanisms for the oxidation of cyclohexanol $(R\!>\!CHOH)$,[49] paraffins (RH)[42,50-53] and acetaldehyde (CH_3CHO)[54-56] are compared on page 15. It may be noted that the first step proposed in each case involves the formation of the respective free radical of the alcohol, paraffin and aldehyde. However in the latter case the following initial reaction also may occur because its activation energy (14 kcal/mole) is low. $RCHO + O_2 \rightarrow RCO' + HO\!-\!O\cdot$. The chain propagation steps 2 to 5, proposed by the researchers remain similar and involve the peroxide and hydroperoxide radicals in each case. The chain termination steps put forth, however differed and hence are given below.

CHAIN TERMINATION REACTIONS FOR CYCLOHEXANOL OXIDATION

$$H^+ + OH^- \rightleftharpoons H_2O$$

CHAIN TERMINATION REACTIONS FOR PARAFFIN OXIDATIONS

(a)   with ketone formation

$$R' + R\!-\!O' \rightarrow RH + R'{=}O$$

$$2R'{=}O \rightarrow ROH + R'{=}O$$

$$2R\!-\!O\!-\!O' \rightarrow ROH + R'{=}O + O_2$$

(b) with formation of aldehydes

$$R-O-O' + M^{2+} \rightarrow RCHO + HO^- + M^{3+}$$

$$H^+ + OH^- \rightleftharpoons H_2O$$

CHAIN TERMINATION REACTIONS FOR ALDEHYDE OXIDATIONS

$$\overset{O}{\overset{\|}{CH_3C'}} + \overset{O}{\overset{\|}{CH_3C}}-O' \rightarrow (CH_3CO)_2O$$

$$\overset{O}{\overset{\|}{CH_3C}}-O' + \overset{O}{\overset{\|}{CH_3C}}-O' \rightarrow (CH_3CO)_2O + O_2$$

$$H^+ + OH^- \rightleftharpoons H_2O$$

Another mechanism which appears to hold for the production of mono- and dicarboxylic acids in the presence of promoters (such as bromine) is considered to involve the following steps.[57-60]

(a) Initiation reactions:

$$RCH_3 + Br' \rightarrow RCH'_2 + HBr$$

$$RCH'_2 + O_2 \rightarrow RCH'_2O-O'$$

$$RCH_3 + RCH_2-O-O' \rightarrow RCH_2-O-OH + RCH'_2$$

(b) Propagation reactions:

$$RCH_2-O-OH + M^{3+} \rightarrow RCH_2-O-O' + H^+ + M^{2+}$$

$$RCH_2-O-OH + M^{2+} \rightarrow RCH_2O' + OH^- + M^{3+}$$

$$RCH_2O-O' + M^{2+} \rightarrow RCHO + OH^- + M^{3+}$$

$$HBr + M^{3+} \rightarrow Br' + M^{2+} + H^+$$

$$RCH_2O' + RCH_3 \rightarrow RCH_2OH + RCH'_2$$

(c) Termination reactions:

$$RCH'_2 + RCH_2O' \rightarrow RCH_3 + RCH=O$$

$$2RCH_2O' \rightarrow RCH_2OH + RCH=O$$

$$RCH_2O-O' + RCH'_2 \rightarrow RCH=O + RCH_2OH$$

$$2RCH_2O-O' \rightarrow RCH_2OH + RCH=O + O_2$$

$$H^+ + OH^- \rightleftharpoons H_2O$$

HOMOLYTIC OXIDATION REACTION MECHANISMS

(Chain initiation and propagation for cyclohexanol, paraffins and acetaldehyde)

1
$$R{>}CH{-}OH \qquad RH \qquad CH_3CHO$$
$$\downarrow M^{3+} \qquad\qquad \downarrow M^{3+} \qquad\qquad \downarrow M^{3+}$$

2
$$RC'OH + H^+ + M^{2+} \qquad R' + H^+ + 2M^{2+} \qquad CH_3\dot{C}{=}O + H^+ + M^{2+}$$
$$\downarrow O_2 \qquad\qquad \downarrow O_2 \qquad\qquad \downarrow O_2$$

$$R{>}C\begin{smallmatrix}O{-}O\cdot\\[4pt]\\OH\end{smallmatrix} \qquad R{-}O{-}O\cdot \qquad CH_3\overset{O}{C}{-}O{-}O\cdot$$

(peroxyradical formation)

3
$$\downarrow R{>}CHOH \qquad\qquad \downarrow RH \qquad\qquad \downarrow CH_3CHO$$

$$R{>}C\overset{OH}{-}O{-}OH \qquad R{-}O{-}OH + R\cdot \qquad CH_3\overset{O}{C}{-}O{-}OH$$
$$+ R{>}C\cdot OH \qquad\qquad\qquad\qquad + CH_3C\cdot{=}O$$

(hydroperoxide radical formation)

$$M^{3+} \qquad\qquad M^{3+} \qquad\qquad M^{3+}$$

4
$$R{>}C\overset{OH}{-}O{-}O\cdot \qquad R{-}O{-}O\cdot \qquad CH_3\overset{O}{C}{-}O{-}O\cdot$$
$$+ M^{2+} + H^+ \qquad + M^{2+} + H^+ \qquad + M^{2+} + H^+$$

5
$$\qquad\qquad M^{2+} \qquad\qquad M^{2+} \qquad\qquad M^{2+}$$

$$RC\begin{smallmatrix}OH\\[4pt]\\O\cdot\end{smallmatrix} + OH^- + M^{3+} \quad RO' + OH^- + M^{3+} \quad CH_3\overset{O}{C}{-}O\cdot + OH^- + M^{3+}$$

$$R{>}CHOH \downarrow \qquad\qquad RH\downarrow$$

6
$$R{>}C{=}O + H_2O \qquad ROH + R\cdot \qquad \text{from 3}$$
$$+ R{>}C\cdot OH$$
$$\qquad\qquad\qquad\qquad\qquad\qquad\qquad \nearrow 2CH_3COOH$$
$$\qquad\qquad\qquad\qquad\qquad\qquad\qquad \searrow (CH_3CO)_2O + H_2O$$

HOMOLYTIC OXIDATION OF OLEFINS

The oxidation of ethylene in the presence of metal ion catalysts appears to follow the same steps as those proposed above for hydrocarbons. In the case of propylene oxidation this reaction mechanism does not account satisfactorily for the formation of secondary products. Bart *et al.* have proposed the following variations in the reaction steps to explain the formation of the secondary products.[61]

1 $CH_2{=}CHCH_2{-}O{-}O' + CH_3CH{=}CH_2 \rightarrow$

$$CH_3CHCH_2OOCH_2CHCH_2$$

2 $CH_3CHCH_2OOCH_2CHCH_2 \rightarrow$

$$\underset{\displaystyle CH_3CH{-}CH_2}{\overset{\displaystyle O}{\diagup\diagdown}} + CH_2CHCH, O'$$

2 (alternate)

$CH_3CHCH_2OOCH_2CHCH_2 + O_2 \rightarrow$

$$\overset{\displaystyle O{-}O'}{\underset{\displaystyle CH_3CHCH_2OOCH_2CHCH_2}{|}}$$

3 $CH_2CHCH_2O' + CH_3CHCH_2 \rightarrow CH_3C'HCH_2OCH_2CHCH_2$

4 $CH_3C'HCH_2OCH_2CHCH_2 \xrightarrow{+O_2} \overset{\displaystyle O{-}O'}{\underset{\displaystyle CH_3CHCH_2{-}OCH_2CHCH_2}{|}}$

Regardless of the paths followed, the intermediate products formed in the initial stages of the reaction may be expected to reach a maximum yield and then decrease as the process continues. Thus the above mechanism suggests that acetone and CO_2 would be formed initially with increasing amounts of propylene oxide, acrolein, acetaldehyde and acids as the oxidation process continues. If it is further extended methanol, ethanol, methyl formate and even propylene glycol may be formed.

Transition metal ion catalysts of the stearate, naphthenate, etc. type contribute significantly to the cost of the final product. Instances are known where the cost of such catalysts accounts for as much as 20 % of the cost of the main and auxiliary materials. One way of reducing these costs is to substitute synthetic organic acids with shorter chains. Another effective measure would be to recover the metal remaining when the catalysts are produced, since the yields generally are quite low, 65-75 %.

Improved catalyst yields also can be achieved by extracting the catalyst with an organic solvent following its synthesis from a solution of the metal salt and the alkali salt of the organic acid.[62] When toluene, *p*-xylene or cyclohexane were employed as extraction solvents solution of $CoCl_2.6H_2O$; $Pb(NO_3)_2$; $NiCl_2.6H_2O$; $FeCl_3.6H_2O$; $FeSO_4.7H_2O$ and $MnCl_4.H_2O$ gave yields up to 95 % with C_7 to C_9 organic acids. The metal ion catalyst solutions contained up to 20 wt % of the salt.

2.1.3   Role of Peroxides and Hydroperoxides

The role of peroxides and hydroperoxides in metal ion catalysis of oxidation reactions has been alluded to in the mechanism discussed earlier. However since these materials are quite stable and can be isolated the following further observations are in order at this point.

Both peroxides and hydroperoxides are found in all oxidations (or auto-oxidations) of hydrocarbons where the CH, CH_2 or CH_3 group is bound to the rest of the molecule by a single bond or where the CH or CH_2 group is bound directly to a carbon atom with a double bond. It is believed that hydroperoxides are produced initially in the normal burning of all hydrocarbons. Methyl and ethyl hydroperoxides have been isolated in the pure state from mercury vapour ultraviolet initiated oxidation of methane and ethane respectively, at 25°C.

Since alkanes and cyclo alkanes oxidise with relative difficulty, the high temperatures required to initiate the reaction cause rapid decomposition of the peroxides and hydroperoxides. Nonetheless the following chemistry appears to hold consistently. In alkanes the tertiary carbon is attacked most readily. In alkenes the hydroperoxide group attacks the CH_2 group next to the double bond or in its absence the CH_3 group next to the double bond. In aromatic compounds the hydroperoxide group enters the side of a branched chain at the CH or CH_2 group next to the nucleus.

The stability of hydroperoxides is generally dependent upon the size of the nucleus. Methyl hydroperoxide is highly explosive whereas ethyl hydroperoxide is more stable although its vapours, also, explode readily when superheated. Tertiary butyl hydroperoxide on the other hand is sufficiently stable that it begins to lose oxygen at about 100°C and explodes only at temperatures of 250°C or greater.

The thermal decomposition of hydroperoxides may proceed according to the following three routes.

HOMOLYTIC DECOMPOSITION

$$R-OOH \rightarrow RO' + OH'$$

In the presence of an alkene the OH' radical may act as an agent of hydrolysis. The formation of acetophenone from cumene hydroperoxide may be explained by this means.

$$C_6H_5C(CH_3)_2-OOH \rightarrow C_6H_5C(CH_3)_2-O' + OH' \rightarrow$$

$$C_6H_5COCH_3 + CH_3OH$$

HETEROLYTIC DECOMPOSITIONS

These may be exemplified by the acid catalysed rearrangement of the Wagner-Meerwein type, of cumene hydroperoxide to phenol and acetone.

$$CH_3-\underset{\underset{CH_3}{|}}{\overset{\overset{C_6H_5}{|}}{C}}-O-OH \xrightarrow{\;H^+\;} -\underset{|}{\overset{|}{C}}-O-\overset{+}{\underset{H}{O}}-H \xrightarrow{\;-H_2O\;} -\underset{|}{\overset{|}{C}}-O^+ \longrightarrow$$

$$CH_3\underset{\underset{CH_3}{|}}{\overset{+}{C}}-O-C_6H_5 \xrightarrow[-H^+]{+H_2O} CH_3-\underset{\underset{CH_3}{|}}{\overset{\overset{OH}{|}}{C}}-OC_6H_5 \longrightarrow$$

$$(CH_3)_2CO + HOC_6H_5$$

ELECTRON TRANSFER INITIATED DECOMPOSITIONS

These initially produce radicals such as

$$R{>}C\overset{\diagup OH}{\underset{\diagdown O'}{}} , RO' \text{ and } RC\overset{\diagup O}{\underset{\diagdown O'}{\parallel}}$$

given in step 5, page 15. It is shown there that these radicals form in the presence of divalent metal ions and react further with the OH^- ion and the respective hydrocarbon in the presence of the trivalent metal ion to produce respectively ketones, alcohols and acids and anhydrides. Note that the alkoxyl radicals act effectively as dehydrogenation agents.

2.2 TECHNOLOGICAL ASPECTS OF HOMOGENEOUS LIQUID PHASE OXIDATION PROCESSES

The pertinent details of the manufacturing processes based on homogeneous liquid phase oxidation reactions are given below under the headings of coordination complex catalysed, metal ion catalysed and noncatalysed reactions. A comprehensive outline of all three types is given here.

SOME TYPICAL HOMOGENEOUS LIQUID PHASE OXIDATION REACTIONS

[Using O_2, air, HNO_3 (#19) catalysts and initiators (I)]

1    $C_2H_4 \xrightarrow[\text{CuCl}_2]{\text{PdCl}_2} CH_3CO_2C_2H_3$

2    $C_2H_4 \xrightarrow[\text{CuCl}_2]{\text{PdCl}_2} CH_3CHO$

3    $CH_3CHO \xrightarrow[\text{acetate}]{\text{Mn}^{2+},\ \text{Co}^{2+}} (CH_3CO)_2O$

4    $CH_3CHO \xrightarrow[\text{acetate}]{\text{Mn}^{2+},\ \text{Co}^{2+}} CH_3COOH$

5    $CH_3COOH \rightarrow (CH_3)_2CO$ (no catalyst)

6    $(CH_3)_2CO \xrightarrow{\text{H}_2\text{O}_2\text{(I)}} iC_3H_7OH$

7    $C_6H_5CH_3 \xrightarrow[\text{Co, Mn acetate}]{\text{CBr}_4\text{(I)}} C_6H_5COOH$

8    $C_6H_5COOH \xrightarrow[\text{Cu(O}_2\text{CC}_6\text{H}_5)_2]{\text{Mg(O}_2\text{CC}_6\text{H}_5)_2\text{(I)}} C_6H_5OH$

9    $(CH_3)_2C{=}CHC_2H_5 \xrightarrow{\text{Co naphthenate}} (CH_3)_2CHCH_2\overset{\overset{\displaystyle O}{\|}}{C}CH_3$

10   $n\text{-}C_4H_8 \rightarrow CH_3COOH$ (no catalyst)

11   $n\text{-}C_4H_8 \xrightarrow[\text{CuCl}_2]{\text{PdCl}_2} C_2H_5CH_3CO$

12   $p\text{-}C_6H_4(CH_3)_2 \xrightarrow[\text{or NH}_4\text{ molybdate}]{\text{Mn, Co acetate}} p\text{-}C_6H_4(COOH)_2$

13   $o\text{-}C_6H_4(CH_3)_2 \xrightarrow{\text{Co, Mn, Cr salts}} o\text{-}C_6H_4\overset{\displaystyle \diagup CH_3}{\diagdown COOH}$

14   $o\text{-}C_6H_4(CH_3)COOH \xrightarrow[\text{or without catalyst}]{\text{CH}_3\text{OH with}} o\text{-}C_6H_4(CH_3)CO_2CH_3$

15   $o\text{-}C_6H_4(CH_3)CO_2CH_3 \xrightarrow[\text{of fatty acids}]{\text{Co, Mn salts}} o\text{-}C_6H_4(CO_2)O$

16   $1,2,4C_6H_3(CH_3)_3 \rightarrow 1,2,4C_6H_3(COOH)_3$ (no catalyst)

17   $\langle\!\langle H \rangle\!\rangle \xrightarrow{\text{Co naphthenate}} \langle\!\langle H \rangle\!\rangle OH + \langle\!\langle H \rangle\!\rangle{=}O$

18 ⬡(H)=O $\xrightarrow[\text{acetates}]{\text{Mn, Ba}}$ $(C_2H_4COOH)_2$

19 ⬡(H)OH + ⬡(H)=O $\xrightarrow[\text{Cu(NO}_3)_2\text{HNO}_3]{\text{NH}_4\text{VO}_3}$ $(C_2H_4COOH)_2$

20 ⬡(H)OH + ⬡(H)=O $\xrightarrow[\text{hexanonitrile (I)}]{\text{azodichloro}}$ $(C_2H_4COOH)_2$

2.2.1 Coordination Complex Catalysed Processes

VINYL ACETATE FROM ETHYLENE

The Hoechst-Uhde Corp process[63-65] is based on the use of $PdCl_2$-$CuCl_2$. The chief reactions may be summarised as follows:

$$2CH_2=CH_2 + CH_3COOH + O_2 \rightarrow$$

$$CH_2=CHOCOCH_3 + CH_3CHO + H_2OCH_3CHO + \tfrac{1}{2}O_2 \rightarrow$$

$$CH_3COOH$$

The process flow diagram is given in Figure 2. The oxidation of the ethylene occurs in reactor 1 in the presence of copper acetate, copper chloride and small amounts of palladium chloride. An important characteristic of the process lies in the facility of varying the molar ratio of acetaldehyde (ALD) to vinyl acetate (VIAC) produced from 0·3:1 to 2·5:1. The control is effected by regulating the flows and the water content of the recycled acetic acid. Water formation from the reaction increases the percentage of acetaldehyde as secondary product[66] while operations at higher pressures favour the formation of vinyl acetate. At 100°C the molar ratios achieved for acetaldehyde to vinyl acetate were 0·6, 1·2, 2·1 and 3·3 respectively, when the catalyst water content was 10, 20, 30 and 40%. The fraction of acetic acid consumed per pass is a function of the temperature, volume of recycled gas and the activity of the catalyst. The conversion of ethylene per pass on the other hand is limited by the allowable concentration of oxygen in the ethylene-oxygen feed mixture. This varies with the operating temperature and pressure, for example at 130°C and 30 atmospheres the upper explosion limit restricts the oxygen content to less than 5·5% of the feed mixture.

The catalyst solution concentrations range between 3 and 6 grams per litre of Cu^{2+} ions and 30-50 milligrams per litre of Pd^{2+} ions. Higher

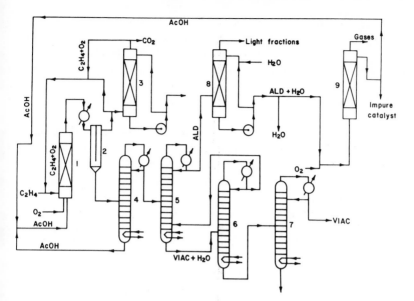

Fig. 2   Hoechst-Uhde process for manufacture of vinyl acetate.
1—reactor; 2—gas-liquid separator; 3—scrubber; 4,5,6,7—distillation columns; 8—scrubber for separation of light fractions from acetaldehyde; 9—reactor; VIAC—vinyl acetate; ALD—acetaldehyde; AcOH—acetic acid.

concentrations of the latter favour increased formation of vinyl acetate and acetaldehyde but also promote dimerisation of ethylene to butene.

The secondary products formed thus include butene and butenyl acetate as well as methyl acetate, glycol diacetate, oxalic acid, formic acid and various chlorinated products. Between 3 and 7% of the ethylene is oxidised completely to CO_2 during the processing with recycle.

The acetic acid introduced to the primary reactor 1 may be obtained by oxidation of acetaldehyde in the catalytic oxidation reactor 9 or from the bottom fraction of the separating column 4 used to distil off the acetaldehyde and vinyl acetate.[67-71]

In the temperature range 100-130°C the reacting masses are kept in the liquid state at pressures of 30-35 atmospheres. Methyl and ethyl chlorides which are formed as byproducts are recycled upon recovery from the aldehyde column. The yield achieved under this condition is about 90%. The catalyst role and regeneration mechanism has been described under Section 2.1.1.

Comparing this process with the basic process based on acetylene the following differences are apparent. For capacities of the order of 50 000 tons per year the investment plant cost for the process based on ethylene is somewhat greater than that based on acetylene. Thus the economic advantages rest primarily with the price difference between ethylene and acetylene. The ethylene process is optimised economically when the molar ratio of acetaldehyde to vinyl acetate has a value of about 1·14.

Vinyl acetate is produced, also, by heterogeneous catalysis from ethylene, acetic acid and oxygen in the gaseous phase. The reactor temperature and pressure ranges in this case are 100-200°C and 1-7 atmospheres respectively. Palladium or a mixture of noble metals on alumina or silica supports have high activity and selectivity—91-94 moles of vinyl acetate and 1 mole of acetaldehyde per 100 moles of ethylene is a typical yield. The conversions per pass vary from 10 to 15% for ethylene and 15 to 30% for acetic acid to 60 to 90% for the oxygen in the feed. Thus it appears that the heterogeneous process based on $PdCl_2CuCl_2$ complex catalyst may be superior to the former processes. However, it is dependent upon the availability of large amounts of acetic acid at low cost. In the final instance the choice of process may be determined by several other interrelated factors of a technical-economical nature.

ACETALDEHYDE FROM ETHYLENE

The Hoechst-Uhde Corp process just described and the Hoechst-Wacker process described below are most important.[72-76] The primary reactions and mechanisms for these processes have been given under Section 2.1.1. Two major distinctions apply on the industrial scale. Purified ethylene may be oxidised with oxygen in a single step or pure and impure ethylene may be oxidised with air in two steps. The required purity of the acetaldehyde, the quality of the ethylene feed and the availability of low cost oxygen determine which process will be selected.

In the two-step process cuprous chloride results from the reduction of about 20% of the cupric chloride available in the first oxidation tower. It is reoxidised in the second tower and recycled to the first at 100°C and 7-8 atmospheres. The reacting mass is partially gaseous and yields of the order of 95% are obtained.

A typical single step process is outlined schematically in Figure 3.

An important feature of this process is that a water scrubber is employed to condense and extract the acetaldehyde from the unreacted gases. The minimum purity preferred in this process is 99·8% for ethylene and 99·5% for oxygen. The heat release is high, being about 58 kilocalories per mole of acetaldehyde formed, hence considerable cooling is required.

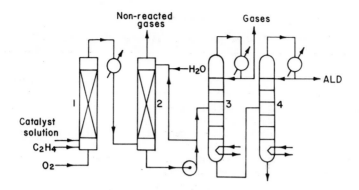

Fig. 3   Hoechst-Uhde process for acetaldehyde from ethylene.
1—reactor; 2—scrubber; 3,4—degasser and distillation column.

Acetaldehyde obtained by (oxidative) dehydrogenation processes, even at high conversion yields, is not competitive with the Hoechst-Wacker process.[77] For example, ethanol may be dehydrogenated over Cu-Cr catalyst at 260-290°C at atmospheric pressure to acetaldehyde with yields approaching 90 % and minor contamination with acetic acid, ethyl acetate and butyl alcohol as byproducts. However, it offers interesting prospects only where the hydrogen coproduct can be utilised effectively in another part of the plant.

Acetaldehyde is obtainable, also, by heterogeneous partial oxidation of ethanol with aid over a silver sieve catalyst at temperatures between 375 and 550°C. With clean preheated air, yields up to 95 % are possible.

Propane and butane when oxidised with preheated air at about 455°C also will produce acetaldehyde with 30-40 % yield, and this process is used when the secondary products formaldehyde, methyl alcohol, higher alcohols, acetone, glycols and acids are desirable end products. The acetaldehyde is recovered from this process by washing the effluent first with a 12-14 % solution of formaldehyde and then with water to extract soluble components prior to azeotropic distillation and final extraction.

The advantages of the Wacker process for producing acetaldehyde can be readily discerned from the data given in Table 2. In 1969 this process accounted for about 50 % of the production in West Germany and in the USA. In Italy a plant to produce 100 000 tons per year of acetaldehyde was put on stream in 1969 and in the same year 350 000 tons of acetaldehyde were produced in Japan by this process.

TABLE 2

Comparison of various processes for manufacturing acetaldehyde

Raw material	Process	Yield (%)	Conversion per pass (%)
ethanol	vapour phase heterogeneous catalytic oxidation	85-95	25-55
ethanol	vapour phase heterogeneous catalytic dehydration	85-95	50
propane or butane	vapour phase heterogeneous non-catalytic oxidation	30-40	25
acetylene	liquid phase homogeneous catalytic hydration	95	50-60
ethylene	liquid phase homogeneous catalytic oxidation	95	95-99

Acetaldehyde is an important raw material for producing large tonnages of chloral, acetic acid, acetic anhydride, peracetic acid, metaldehyde, paraldehyde, polyacetaldehyde, normal butanol, pentaerithnitol, etc.

ACETONE FROM PROPYLENE

It has been stated previously that olefins can be catalytically oxidised in the presence of coordination complexes of the transition metals to produce ketones. The following reactions apply to propylene oxidation (hydration oxidation to be more correct):

$$CH_2{=}CH{-}CH_3 + PdCl_2 + H_2O \rightarrow CH_3COCH_3 + Pd + HCl$$

$$2CuCl_2 + Pd \rightarrow PdCl_2 + 2CuCl$$

$$2CuCl + 2HCl + \tfrac{1}{2}O_2 \rightarrow 2CuCl_2 + H_2O$$

$$CH_2{=}CH{-}CH_3 + \tfrac{1}{2}O_2 \rightarrow CH_3COCH_3 \quad \text{(overall)}$$

Secondary products formed in this process consist primarily of 0·5-1·5% propylaldehyde and about 2% of mono- and dichloro-acetone.

As in the case of ethylene, propylene probably also forms a ligand aquo-complex with the palladium salt $[PdCl_2(OH)C_3H_6]^-$ as an intermediate product and hence the oxidation of the molecule occurs through abstraction of the oxygen from the water molecule. The reaction rate varies directly with the $PdCl_2$ concentration which supports this postulate. Ferric chloride or mixtures of ferric and cuprichloride salts may be used as palladium reoxidation agents. The pH of the reacting solutions is generally maintained between 3 and 4. Since this process is simple, highly selective and gives high conversions and yields, it competes favourably with other processes both at low and high capacities. The Hoechst-Uhde process is outlined schematically in Figure 4. The Hoechst-Uhde process

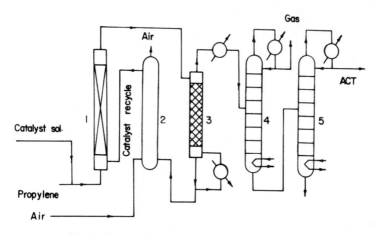

Fig. 4   Hoechst-Uhde process for manufacture of acetone.
1—reactor; 2—catalyst regenerator; 3—stripper; 4,5—degasser and distillation column.

for making acetone from propylene consists of two steps.[78-82] In the first step, the propylene is oxidised with air at about 100°C and 9-12 atmospheres pressure, while in the second step the cuprous chloride is re-oxidised and the palladium chloride catalyst is regenerated for recycle. The heat of reaction is quite high, 61 kilocalories per mole, and the oxygen conversion per pass is 100 %. Thus the residual gases may be recovered and employed as inert gas. Yields are of the order of 93 %. The propylene feed should be at least 90 % pure.

METHYL ETHYL KETONE FROM BUTENE

Another practical example of homogeneous liquid phase catalytic oxidation using coordination complexes of the $PdCl_2CuCl_2$ system is that of the Hoechst-Uhde Corp for producing methyl ethyl ketone from butene.[83,84] The schematic flow diagram for this process is given in Figure 5.

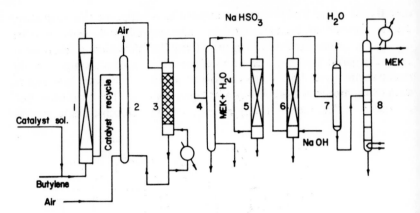

Fig. 5   Hoechst-Uhde process for the manufacture of methyl ethyl ketone.
1—reactor; 2—catalyst regenerator; 3—stripper; 4—separator for mixture of H_2O—MEK; 5—extractor with $NaHSO_3$; 6—extractor with $NaOH$; 7—evaporator; 8—distillation column; MEK—methyl ethyl ketone.

Butene like propylene and ethylene undergoes the same type of reactions with the $PdCl_2$-$CuCl_2$ catalyst system. The oxidation of the butene is carried out at about 100°C and 10-15 atmospheres pressure with air being supplied to the catalyst reoxidation or regeneration tower. Again the heat of reaction is high (60 kilocalories per mole of ketone formed) and the butene should have a purity of at least 90%. The yields from this raw material however are somewhat lower but do not fall below 80% under the most adverse operating conditions.

2.2.2   Oxidations with Metal Ion Type Catalysts

ACETIC ANHYDRIDE FROM ACETALDEHYDE

The Shawinigan Process[85] is based on the following sequence of reactions:

$$CH_3CHO + O_2 \longrightarrow CH_3COOOH \xrightarrow{CH_3CHO} (CH_3CO)_2O$$

As outlined earlier the formation of peracetic acid (a hydroperoxide) is promoted by the presence of manganese, cobalt, copper, etc. acetates. Its further reaction with acetaldehyde to form the anhydride also is catalysed by these metal salts. Improved yields of the anhydride are obtained by maintaining low reaction temperatures which reduces the hydration rate without seriously reducing the oxidation rate if the catalyst concentration and activity are maintained at a high level. Further yield improvements are effected commercially by the use of continuous flow reactors and separators to isolate the product as it is formed. With oxygen or air at 60-80°C and 4-7 atmospheres pressure the yield under continuous operation is about 95%. The use of solvents such as benzene or alkyl acetates is reported to permit higher levels of peracetic acid to be maintained in the reaction mixture without danger of explosion and to facilitate the rapid removal of water by azeotropic distillation.[86-88]

ACETIC ACID AND OTHER ACIDS FROM ACETALDEHYDE OR ALKANES

Acetic Acid from Acetaldehyde

As can be seen from Section 2.2 at the intermediate or less drastic oxidation stage of acetaldehyde, acetic acid is formed *in lieu* of its anhydride. In the Shawinigan process for continuous acid production the aldehyde is premixed with the catalyst solution before being fed to the base of the oxidising reactor.[89,90] Air (or oxygen) is fed at different levels to the reactor which is maintained at 70-80°C and about 1 atmosphere pressure. The conversion per pass is about 50% and the yield is about 95%.

When the reaction is carried out in a batch process somewhat lower temperatures and higher pressures (about 5 atmospheres) are employed to effect the oxidation in a medium of acetic acid. The batch yields obtained also are about 95%.

Acetic Acid from Alkanes

In the Celanese process[91-93] the primary feed is comprised largely of normal butane and the overall reaction approximates

$$n\text{-}C_4H_{10} + \tfrac{5}{2}O_2 \rightarrow 2CH_3COOH + H_2O$$

The use of higher alkanes (C_4-C_7) and the presence of branched chains reduces yields of acetic acid. The chief secondary products apart from formic acid are methyl ethyl ketone and various alkyl acetates.[94] The nature and yield of the various products are determined largely by the type and selectivity of the catalyst employed. By proper choice the induction period of the reaction may be reduced or eliminated and the degree of hyperoxidation controlled.[95]

Cobalt acetate catalyst favours the formation of acetic over formic acid while manganese acetate is somewhat less selective. At 170-180°C and 55 atmospheres pressure the products formed with these two catalysts were found to contain respectively 76% CH_3COOH with 6% $HCOOH$

and 61 % CH_3COOH with 25 % HCOOH. Both catalysts have a marked effect on the induction time.

Hyperoxidation was considerably reduced by reducing the conversion per pass to less than 30 %, by adding certain inhibitors and by maintaining the oxygen concentration well below the explosion limits.[96] The usual ratio of air to butane by weight is about 5 to 1.

Acetic Acid from n-Hexane

Recent studies[97] have shown that about 1·5 moles of acetic acid can be obtained per mole of n-hexane when the latter is oxidised using manganese naphthenate as catalyst. In the presence of anhydrous sodium carbonate or calcium oxide in superfine form the conversion of n-hexane was reduced as was the quantity of formic acid byproduct. Larger quantities of propionic and butyric acid were formed, however, when these additives were employed. In each case the reduction was carried out at 160°C with a partial pressure of oxygen in the reactor of 15 atmospheres and the liquid products were recovered by cooling the effluent to −10°C.

These studies indicate that the primary products of n-hexane oxidation are the higher acids whose further oxidation or decarboxylation is diminished by the formation of insoluble salts in the presence of the sodium carbonate or calcium oxide powders.

Using 0·64 grams of manganese naphthenate and 100 grams of pure hexane, 18 grams of formic acid, 84 grams of acetic acid, 20 grams of propionic acid and 6 grams of butyric acid were recovered. This is a significantly high yield.

These studies indicate that the oxidation rates of carbon atoms 2, 3 and 3, 4 are similar while the oxidation rates of the atoms 1, 2 are about 5 times smaller. The oxidative decarboxylation of the higher acids proceeds at about the same rate but is greatly reduced for acetic acid because of its greater resistance to oxidation.

The following kinetic equations were proposed to account for the oxidation of n-hexane in the liquid phase.

$$\frac{-d(RH)}{dt} = K(RH)$$

$$\frac{d(C_4)}{dt} = 0\cdot5K(RH) - K_d(C_4)$$

$$\frac{d(C_3)}{dt} = 0\cdot5K(RH) - K_d(C_3) + K_d(C_4)$$

$$\frac{d(C_2)}{dt} = 2 \cdot 5K(RH) + K_d(C_3)$$

where

K = oxidative rate constant of hexane

K_d = oxidative decarboxylation rate constant of acids

(RH) = concentration of n-hexane

(C_2) = concentration of acetic acid

(C_3) = concentration of propionic acid

(C_4) = concentration of butyric acid

Fatty Acids from Higher Alkanes

The information on the oxidation of higher alkanes is abundant and only certain aspects will be discussed here. The fatty acids of major interest to the soap industry are produced by oxidative splitting of the C_{18}-C_{30} alkanes.[98] The molecular weights of these products range between 250 and 450. Their melting points range from 28 to 66°C. The prime reaction may be best expressed by the following equation:

$$R_1(CH_2)_nCH_2-CH_2(CH_2)_nR_2 \xrightarrow{\frac{5}{2}O_2}$$
$$R_1(CH_2)_nCOOH + R_2(CH_2)_nCOOH + H_2O$$

In the Deutsche Hydrierwerk-Rodleben process, 40-60 m³ of air per ton of hydrocarbon per hour are fed to the oxidising reactors which are maintained at 105-120°C and 14-62 atmospheres. Manganese salts are used as catalysts. About 60 % of the raw material is converted to fatty acids suitable for soap manufacture, up to 25 % is converted to shorter fatty acids and about 10 % is oxidised to CO_2 and CO.

Russian researchers have reported that the yield of C_7-C_9 fatty acids from solid paraffins can be increased from a level of 8-10 % to about 40 % by adding 30 % liquid paraffin to the solid.[99] In this study four paraffin fractions with boiling ranges between 260-350°C (Type A), 275-320°C (Type B), 250-300°C (Type C) and 230-270°C (Type D) were evaluated. The oxidations were carried out batchwise using 90 kilograms of paraffin and 21 litres of air/min/kg of paraffin. For the paraffins A, B and C the reaction temperature ranged between 107 and 125°C, for type D a constant reaction temperature of 120°C is cited. The catalysts were prepared by reacting manganese and sodium salts with C_5-C_9 acids and the reaction time was varied between 12 and 20 hours. The raw material consisted of $\frac{1}{3}$ of fresh paraffin (one of the above four types) and $\frac{2}{3}$ nonsaponifiables. Paraffin conversions of 47-50% were achieved. The yield of fatty acid products decreased from about 99% for type A diluent to 93% for type D. The yield of water soluble acids increased in the opposite manner rising from 5-6% for type A to about 15% for type D. The yield of C_7-C_9 acids increased also as the molecular weight of the fresh paraffin was decreased reaching the maximum for type D (about 42%). This portion was concentrated readily by distillation to 90-95% fatty acid content.

BENZOIC ACID FOR TOLUENE

The Mid-Century process oxidises toluene to benzoic acid with air at about 200°C and 30 atmospheres pressure using a bromine compound as an initiator and cobalt or manganese acetate as the catalyst. The yield obtained is about 90%. The overall reaction is:

$$C_6H_5CH_3 + \tfrac{3}{2}O_2 \rightarrow C_6H_5COOH + H_2O$$

The heat of reaction is abstracted by vapourising, condensing and recycling the excess toluene in the reactor.

In the SNIA-Viscosa process the reaction temperatures (150-170°C) and pressure (~ 10 atmospheres) are both lower than those utilised in the Mid-Century.[103-105] Moreover, no initiator is used with the cobalt acetate catalyst whose concentration is held between 100 and 150 ppm. The schematic representation of this process is given in Figure 6.

In this process benzylic alcohol and benzaldehyde formed as intermediate oxidation products are recovered from the stripper 2 and distillation column 3 and recycled to the oxidation reactor 1. The resultant benzoic acid has a minimum purity of 99%.

The problem of obtaining benzoic acid of high purity at optimum yield has been studied thoroughly over several years[106,107] and many schemes have been tested.

The recrystallisation of the acid from water solutions is faced with the problem that at atmospheric pressure its solubility is low (about 5% at 95°C) hence large volumes of water are required with the attendant high demand for energy to mix and filter or centrifuge the solutions. At temperatures above the critical (117°C), benzoic acid is very soluble in water but recrystallisations under pressure pose operational difficulties too. On cooling three layers are formed. One of these, containing 87% benzoic acid at 101·4°C, forms a monolithic block on further cooling thereby

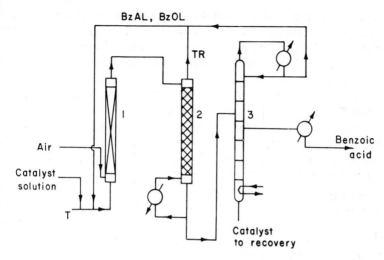

Fig. 6   SNIA-Viscosa process for benzoic acid production.
1—oxidation reactor; 2—stripper; 3—column; T—toluene; R—recovery; BZAL—benzaldehyde; BZOL—benzylic alcohol.

seriously impairing the operation of the crystalliser. A further complication arises in that the remaining benzoic acid is not freed of resins and can be recovered only in impure form. Additional problems and costs arise during drying because sublimation losses are high.

Although the solubility of benzoic acid in toluene is significantly higher than in water (about 12 % at 20°C) its recrystallisation from this medium is equally cumbersome. The recovery must be made at low temperatures and although drying can be effected at lower temperatures than in the case of water the sublimation losses nonetheless are still appreciable.

The modern processes which employ distillation towers to fractionate out the intermediate and secondary reaction products appear to offer the best means for producing benzoic acid of high purity. The products formed from oxidation of toluene yield four major fractions (a) the toluene fraction (b) a highly volatile fraction comprised primarily of benzylic alcohol, benzaldehyde and benzyl acetate (c) a less volatile fraction (benzyl benzoate, etc.) and (d) the benzoic acid fraction which contains diphenyl, diphenyl methane, etc.

By analysing the equilibrium data for toluene-benzaldehyde, benzalde-hyde-benzoic acid, benzyl alcohol-benzoic acid, benzyl acetate-benzoic acid, benzoic acid—benzyl benzoate and benzaldehyde-benzyl acetate mixtures and taking into account the effectiveness of rectification columns it has been calculated that 7-9 trays in a column suffice to give a high separation of the impurities from benzoic acid. Moreover it has been established that the amount of resins formed in 10 hours from benzoic acid drops from 3·7% at 250°C, to 1·7% at 200°C to a negligible amount at 180°C.[108] Hence the modern fractionation and purification steps are carried out using four columns as shown in Figure 7.

In the first column 1 the distillation is conducted at atmospheric pressure with a maximum temperature of 180°C. The bottom fraction from this column is fed to the evaporator(s). The residue from the latter is used in the regenerator of the catalyst while the top fraction is fed back to column 2. This column operates at 30-40 mm Hg pressure and 150-160°C but for the top where the temperature is about 200°C yielding a fraction which is recycled to the reactor. The residue from column 2 is fed to column 3 where it is distilled at 20-30 mm Hg pressure and 147-156°C. The top fraction is benzoic acid which is collected while the bottom fraction or residue is recycled to the reactor. By recycling the intermediate fractions yields of 97-98% are obtained.

PHENOL BY THE TOLUENE-BENZOIC ACID ROUTE

The Dow Chemical Co. process as improved by Montecatini[109-112] is based on the following overall reactions:

$$C_6H_5CH_3 + \tfrac{3}{2}O_2 \rightarrow C_6H_5COOH + H_2O$$
$$C_6H_5COOH + \tfrac{1}{2}O_2 \rightarrow C_6H_5OH + CO_2$$

In Figure 8 only the second phase oxidation is depicted schematically.

The products from the first phase oxidation are fed to the first distillation column 1 where the benzoic acid is separated from other components and fed to the second phase oxidation reactor. Magnesium benzoate is employed here as the initiator and copper benzoate as the catalyst. The evidence suggests that cupric ions function as the oxidising agent since cupric benzoate is reduced by benzoic acid to cuprous benzoate and the latter is reoxidised to the cupric state by air.[113-115]

The products from the second phase oxidation reactor are distilled twice to produce pure phenol. The heavy fraction from column 2 is primarily benzoic acid and is recycled to the reactor. The latter is maintained at 220-245°C and one atmosphere of pressure. The ratio by weight of water to phenol in the reactor is about 2:5 while the volumetric ratio of inert gas plus steam to phenol is approximately 20:1. These conditions

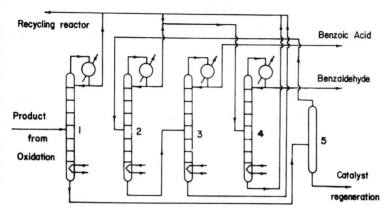

Fig. 7   Benzoic acid recovery and purification process.
1,2,3,4—rectifying columns; 5—evaporator.

reduce but do not eliminate tar formation in this reactor. These tars form the heavy fractions from the reactor and are removed from the process by extraction in column 6. The yield of phenol obtained is about 90%.

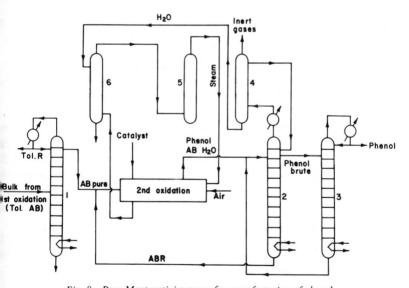

Fig. 8   Dow-Montecatini process for manufacturing of phenol.
1,2,3—distillation columns; 4—decantor; 5—evaporator; 6—extractor; Tol—toluene; R—recycle; AB—benzoic acid.

The large number of factors which must be considered make quantitative economic evaluation of the various methods for manufacturing phenol rather difficult.

The classical route, batch or continuous, is based on sulphonation of benzene followed by fusion with alkali. The overall yield lies between 85 and 90%. Initial plant investments are low but operating costs due to heat requirements are appreciable. The selling price is about $323·00 per ton of phenol.

The Dow process which follows the chlorination of benzene by an alkali hydrolysis produces phenol with an overall yield of 80-95%. The chlorine for the first step is obtained by electrolysis. Thus the economics of this process depends upon the cost of chlorine and sodium hydroxide, as well as the value of secondary products (diphenyl oxide and chloro-benzene) and of electrical energy. Selling price is about $384·00 per ton of phenol.

The Raschig process, which involves vapour phase chlorination of benzene, vapour phase hydrolysis of the latter with steam and oxidation of the hydrochloric gas with air to regenerate the chlorine, produces phenol with yields of 72-79%. The economic value of this process depends on the value of the secondary product, dichlorobenzene and on the capacity of the installation. The selling price is about $227·00 per ton of phenol.

The cumene process, in which benzene is alkylated to form cumene, the cumene oxidised to the peroxide and the latter decomposed to yield acetone and phenol, gives overall yields ranging between 70 and 90%. Its economic value depends primarily on the value of the co-product acetone. The amount of the latter ranges from 0·6 to 0·66 parts per part of phenol. The selling price is about $227·00 per ton of phenol.

Since the selling price of phenol from the Dow-Montecatini process is also about $227·00 per ton a choice between it and either the Raschig or Cumene process will be determined by factors such as availability of raw materials, marketability or other use of secondary products, initial investment requirements and other related considerations.

PROPYLENE OXIDE FROM PROPYLENE IN ORGANIC MEDIA

It has been stated previously that the transition metals, their oxides and anion salts also function as oxidation catalysts. The Escambia process[116-119] for producing propylene oxide from propylene as an example, employs copper or copper oxide or various salts of vanadium, chromium, cobalt, etc. as the catalyst to effect yields as high as 90%. The reactions in this case are carried out in organic solvent media such as benzene or an alkyl acetate at temperatures ranging between 130 and 300°C with pressures close to 45 atmospheres.

CYCLOHEXANOL AND CYCLOHEXANONE FROM CYCLOHEXANE

In Section 2.2 the oxidation of cyclohexanol (CHOL) and cyclohexanone (CHON) to adipic acid (ADP) was included for completeness. References[120-131] describe the characteristics and give details on these oxidations.

In the Dupont process cyclohexane (CHA) is oxidised with air (about 170 litres/kg CHA) at 150-160°C and 10-10·5 atmospheres pressure in the presence of cobalt naphthenate as catalyst. The conversion per pass is about 10% while the selectivity on a continuous basis is about 63%.[132] The low conversion insures a modest oxidation rate which reduces the degree and amount of over oxidation.

When boric acid (H_3BO_3) is used as an additive to the process it functions both as a promoter and an esterification agent with a consequent improvement in both conversion and selectivity of yield. In the presence of boric acid, 90-95% of the CHA is converted to CHOL and CHON as compared with 75% or less in other processes. Moreover because of the marked effect on selectivity the CHOL to CHON ratio in the product is altered from the usual 1:1 to 9 or 10:1. This results from the stabilisation of the CHOL intermediate by esterification.[133,134] Another important control parameter of this process is the partial pressure of water. If this is kept low (by removing the water formed during the reaction) the yield of CHOL and CHON is improved.

The use of boric acid as an additive has been extended to the manufacture of alcohols and ketones from paraffins, bifunctional compounds from cyclohexamine, aniline from cyclohexylamine, etc.

A schematic diagram of the Scientific Design Co. process for producing CHOL and CHON from CHA in the presence of boric acid is given in Figure 9.[135]

It is believed that boric acid reacts with the intermediate cyclohexyl hydroperoxide to produce peroxy borates with subsequent formation of the cyclohexyl borate ester. The acid is recovered from this and other esters by hydrolysis in column 3 and recycled to the oxidation reactor 2. Part of the unreacted CHA and all of the CHA recovered from the distillation column 5 is returned to the CHA reservoir 1 for recycle. The mixture of CHOL and CHON obtained from the base of column 5 may be used directly for manufacture of adipic acid.

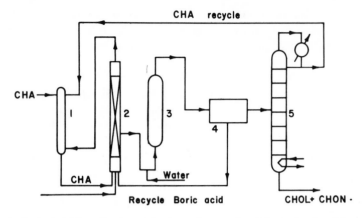

Fig. 9   Scientific Design Co. process for the production of cyclohexanol and cyclohexanone. 1—cyclohexane storage vessel; 2—oxidation reactor; 3—hydrolyser; 4—boric acid recovery unit; 5—distillation column; CHA—cyclohexane, CHOL—cyclohexanol, CHON—cyclohexanone.

The oxidation of mixtures of CHOL and CHON or of CHON is carried out in one of two ways. In the first method 1 part of the CHOL and CHON mixture is reacted with five parts of 50% HNO_3 in the presence of a catalyst comprised of a mixture of $Cu(NO_3)_2$ and NH_4VO_3 as in the DuPont process.[136-138] The dual catalyst permits yields of about 90% to be obtained whereas with ammonium metavanadate alone the yield of ADP is only about 75%.

The oxidation reaction is initiated by the nitrogen oxides which exist in equilibrium with nitric acid. At 25°C about 10% of the total oxides exist in the form N_2O_3 while the remainder is found in form of NO and NO_2. Both of the latter compounds have the capacity (because of their structure) to form free radicals upon contact with hydrocarbons. The NO and NO_2 molecules are free radicals in the sense that one of their orbitals is occupied by a single electron. Each molecule therefore tends to dimerise or to accept an electron from a hydrocarbon transforming the latter into a free radical and thereby promoting homolytic organic reactions. The tendency of these molecules to associate with each other and thereby satisfy the coupling requirements of their respective single electrons is not great in nitric acid solution as demonstrated by the equilibrium data at 25°C given above.

During oxidation of CHOL and CHON to adipic acid the concentration of nitric oxides in solution increases thereby favouring the reactions. It is necessary however to pressurise the system to about 2 atmospheres pressure to reduce losses and improve recovery of NO and NO_2. Conversion of these to N_2O leads to non-recoverable loss of nitric acid.

Because the oxidation of CHOL and CHON with HNO_3 is most important in relation to other processes intensive kinetic studies have been made to help define optimal operating conditions. For example it has been established that the oxidation of CHOL with HNO_3 to adipic acid occurs in two steps. In the first step, nitrocarboxylic acid (NCA) is formed and in the second the NCA is converted to ADP. The following reaction equations were shown to apply[139]

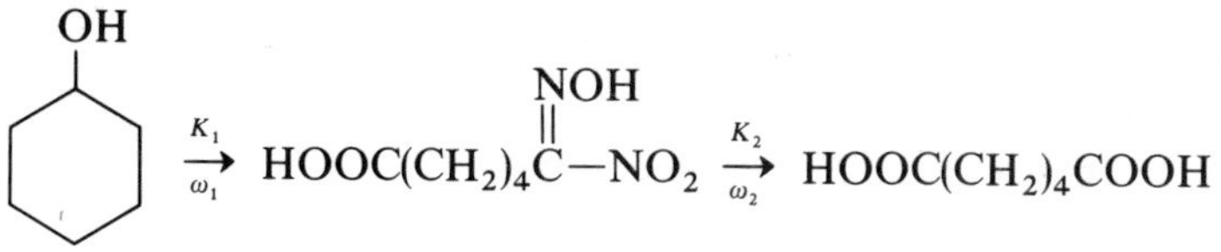

where K_1 and K_2 are first order reaction rate constants and ω_1 and ω_2 are the respective rates of the reactions per unit volume.

The rate of formation of NCA in a continuous flow reactor given by the differential equation,

$$\frac{d(CV)}{dt} = vC_0 + V\omega_1 - V\omega_2 - vC_t \tag{1}$$

where

C_0 = initial concentration of NCA

C_t = concentration of NCA in the reacting mass at time t

V = volume of reactor

v = feed rate of reactant to the reactor (litres per min.)

reduces steady state to,

$$\omega_1 - \omega_2 - C_t/\tau = 0 \quad \text{when } C_0 \text{ is zero} \tag{2}$$

since V/v represents the contact time, τ. Thus the rate of NCA formation was calculated from the formulas

$$C_t = \omega_1\tau$$
$$K_1 = \omega_1\sqrt{(C_\infty - C_t)} \tag{3}$$

where C_∞ represents the theoretical concentration of NCA that would be established if the reaction proceeded only to NCA. Data for these calculations were obtained by varying the reaction time (from 1·9 to 24·8 minutes) by adjusting the feed rate of CHOL and 57% HNO_3 (in a molar ratio of 1:7), and determining the concentration of NCA. The latter was found to range between 0·48 and 0·55 moles per litre. Using equation 3, K_1 was found to remain virtually unchanged for 5-8 fold changes in contact time. At 30°C the value was about 0·5 min^{-1} and at 35°C, 0·86 min^{-1}.

Thus having established that the first step in the oxidation of CHOL with HNO_3 proceeds via a first order reaction these authors then tested the further assumption that the second step also is first order by determining the yield of NCA and the K values over the temperature range 10-60°C using the following modified form of equation 2:

$$C_\infty - \frac{1}{K_1}\left[\frac{C}{\tau}(1 + K_2\tau)\right] - C(1 + K_2\tau) = 0$$

In this way they established that the maximum yield for the first phase oxidation of CHOL (to NCA) occurs between 45 and 55°C.

The second method of producing adipic acid from cyclohexanol and cyclohexanone is by converting the former to the latter (or separating it out) and oxidising the latter with air. In the I.G. Farben process[140,141] CHON is dissolved in an organic acid solution containing a mixture of Ba^{2+} and Mn^{2+} ion catalysts (in the ratio 2·5 to 1 by weight) where it is oxidised with air at 80°C and one atmosphere pressure. The spatial volumetric flow rate of air is about 275 litres per litre of CHON solution per hour. The maximum conversion per pass is about 45% and overall yields in excess of 90% are achieved. In an improved modification of the Farben process mixtures of CHOL and CHON are employed. The operating conditions remain about the same but azodichlorohexano nitrile is added as an initiator of free radical reactions.[142]

METHYL ISOBUTYL KETONE FROM METHYL PENTENE

In the British Petroleum process[143] the oxidation of methyl pentene to methyl isobutyl ketone is effected at temperatures between 45 and 75°C at 10-11 atmospheres pressure in the presence of cobalt naphthenate as catalyst. The yield in accord with the following reaction is about 90%:

$$(CH_3)_2C=CHC_2H_5 + O_2 \rightarrow (CH_3)_2C=CH\overset{\overset{\displaystyle O}{\|}}{C}-CH_3 + H_2O$$

$$\downarrow$$

$$+ H_2$$

$$\downarrow$$

$$(CH_3)_2CHCH_2\underset{\underset{\displaystyle O}{\|}}{C}-CH_3$$

PRODUCTION OF TEREPHTHALIC AND o-PHTHALIC ACIDS

Terephthalic Acid from p-Xylene

In the Mid-Century process terephthalic acid is produced by oxidising p-xylene with air using acetic acid as the solvent and cobalt or manganese acetate or ammonium molybdate as catalyst. Sodium bromide also is added to the solution to act as an initiator to reduce the induction period. The concentration of p-xylene in solution is of the order of 30 wt%. At 195-205°C and 28-30 atmospheres pressure yields of the order of 90% are obtained.[144-147]

The Teijin process differs from the above in that the reaction is carried out at 100-130°C and 10 atmospheres pressure.[148,149] This is achieved by the use of a cobalt base catalyst which is added in an amount approximating 0.1% of the terephthalic acid produced. The yield obtained at these milder conditions is appreciably higher, too. The consumption of p-xylene and acetic acid being about 65 and 100 kilograms respectively per 100 kilograms of acid produced. This process is depicted schematically in Figure 10. Note that in this process the acetic acid used as a solvent both for the reaction and for the crystallisation purification of the acid is continuously purified by stripping and distillation. The catalyst also is continuously treated and recylced in purified form. In these treatments the intermediate products of oxidations are removed by recycle thereby pre-purifying the acid product; the main purification of the latter is achieved by crystallisation from purified acetic acid.

Typical kinetic curves for the formation of p-toluic aldehyde (1), p-toluic acid (2) and terephthalic acid (3) are given in Figure 11. The data for these curves were obtained with the aid of gas liquid chromatography and apply to a cobalt bromide (cobalt acetate and sodium bromide) catalysed system.[150]

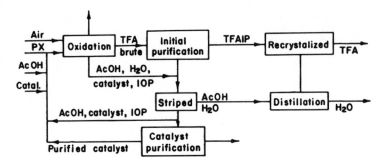

Fig. 10   Teijin process for manufacture of terephthalic acid.
PX—p-Xylene; ACOH—acetic acid; IOP—intermediate oxidation products; TFA—terephthalic acid; TFAIP—terephthalic acid, initially purified; PP—prepurification.

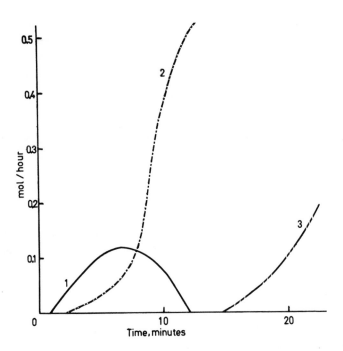

Fig. 11   Kinetic curves for oxidation of p-xylene.
1—Rate of formation of p-toluene aldehyde; 2—Rate of formation of p-toluic acid; 3—Rate of formation of terephthalic acid (temperature—105°C, concentrations (mols./l): for PX—0·56%, for water—6·26 × 10^{-2}, for CoAC$_2$—5·43 × 10^{-2}).

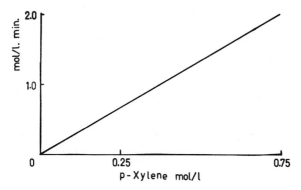

Fig. 12   Rate of formation of p-toluic acid from p-xylene as a function of p-xylene concentration. (Temperature: 70°C; concentration (moles/l); for NaBr—6·3 × 10⁻²; for CoAc₂— 7·3 × 10⁻².)

The curves of Figure 11 indicate that terephthalic acid is not formed until *p*-toluic aldehyde has been completely converted to *p*-toluic acid and the concentration of the latter has reached a maximum. The rate of *p*-toluic acid formation is first order with respect to the concentration of *p*-xylene and of cobalt and bromine ions as shown by curves in Figures 12, 13 and 14. It is independent of oxygen partial pressures from 0·2 to 1·0 atmosphere.

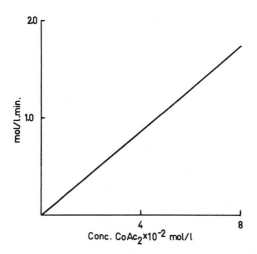

Fig. 13   Rate of formation of p-toluic acid from p-xylene as a function of cobalt catalyst concentration.

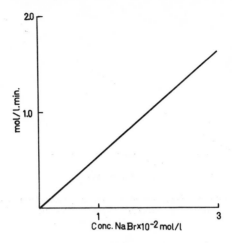

Fig. 14   Rate of formation of p-*toluic acid from* p-*xylene as a function of bromide activator*
concentration.
(Temperature: 70°C; concentrations (moles/l); for $CoAc_2$—7·3 $\times$ 10^{-2}; *for PX—0·515.)*

The rate of formation of terephthalic acid from *p*-toluic acid also is first order with respect to the concentration of cobalt ions as shown in Figure 15. Unlike the primary oxidation of *p*-xylene, the oxidation of this intermediate product varies proportionally with the square root of the oxygen pressure as depicted in Figure 16.

With regard to bromine ions this oxidation step is not influenced by concentration changes, although it does not occur when bromine ions are totally absent. When the cobalt catalyst is replaced by either a nickel,

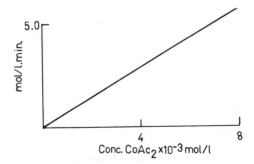

Fig. 15   Rate of formation of terephthalic acid from p-*toluic acid as a function of cobalt*
catalyst concentration.
(Temperature 90°C; concentrations (mol/l); for NaBr—7·2 $\times$ 10^{-2}; for p-*toluic acid—0·46.)*

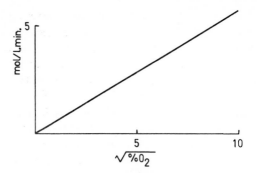

Fig. 16   Rate of formation of terephthalic acid from p-toluic acid as a function of oxygen concentration.
(Temperature 90°C, concentrations (mol/l); for NaBr—7·2 × 10⁻²; p-Toluic acid—0·46; CoAc₂—9·4 × 10⁻².)

chromium or manganese catalyst, no oxidation of p-toluic acid takes place even after 150 minutes with sodium bromide added as activator. With the cobalt catalyst a conversion of about 25% is achieved in 30 minutes. In the case of mixed metal ion catalysts (i.e. Co with Ni, Cr or Mn ions) the best results are obtained when small amounts of nickel salts are co-added as shown in Figure 17.

The oxidation of p-xylene to p-toluic acid can be expressed in the form of equation 1 below; that of p-toluic acid to terephthalic acid by equation 2.

$$\omega_1 = K_1[Co^{2+}][Br^-][C_8H_{10}] \tag{1}$$

$$\omega_2 = K_2[Co^{2+}][O_2]^{0.5} \tag{2}$$

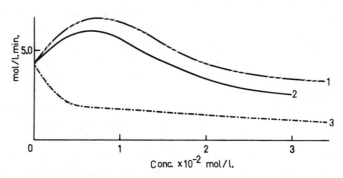

Fig. 17   Rate of formation of terephthalic acid from p-toluic acid as a function of mixed metal salt catalyst concentration.
Curve 1—Ni salt as co-additive; 2—Cr salt as co-additive; 3—Mn salt as co-additive.
(Concentrations (mol/l): for CoAc₂—6·8 × 10⁻²; for NaBr—7·2 × 10⁻²; p-toluic acid.)

Some kinetic comparisons also have been made between cation catalysts such as Co^{2+}, Co^{3+} and Cr^{3+} which exchange electrons rapidly and others such as VO^{2+} which exchange electrons at a slow rate.[151] Under isothermal conditions it was noted that the time required to form the active complex which catalyses the formation of p-toluic acid is determined by the nature of the cation. For Co^{3+}, Co^{2+}, Cr^{3+} and VO^{2+} the respective times noted were 1, 17, 20 and 60 minutes. As might be expected the formation of the cation-ligand complex is slowest for the case of the vanadyl ion. The nature of the organic anion component of the catalyst also exerts a small influence in this regard due to the fact that it determines the solubility of the catalyst in the hydrocarbon.

The activities of the catalysts with respect to the formation of p-toluic acid increase in the order

$$VO^{2+} \rightarrow Co^{2+} \rightarrow Co^{3+} \rightarrow Cr^{3+}$$

For example at catalyst concentrations of 5.6×10^{-3} gram cations per litre and 1 litre air per minute the yields of p-toluic acid from p-xylene at 115°C, respectively were 1·75, 8, 10·4 and 11 % for the above order of cations. Under the same conditions the greatest anion effect on yield was noted between acetylacetate (11·3 %) and naphthenate (8 %). The yield of p-toluic acid also varies with the catalyst concentration. The optimum concentration expressed in gram cations per litre of p-xylene was found to be about 2×10^{-2} for chromium caprylate and 1·3 for cobalt caprylate.

Under the test conditions infra-red spectrophotometric measurements showed that the p-toluic aldehyde concentration increases during the early part of the reaction, then remains at a constant level. The concentration of p-toluic alcohol (as determined by gas chromatography on an Apiezon-L column) passed through a maximum at about 2 hours for all catalysts. Corresponding to the catalysts' activity, the lowest concentration of the alcohol occurred for the catalyst with the highest conversions or yields of p-toluic aldehyde and p-toluic acid.

In Figure 18 the time yield of p-toluic acid is graphed with that of the alcohol, aldehyde and ester of p-toluic acid for a cobalt acetylacetate catalysed oxidation of p-xylene.

The role of the solvent in the liquid phase oxidation of p-xylene has been studied quite extensively also.[152] Acetic, propionic, n-butyric, n-valeric, isovaleric, trimethyl acetic, caprionic, caprylic and oenanthic acids were tested along with chloro-, dichloro- and bromobenzene. It was found that the induction period was decreased and the reaction rate increased as the

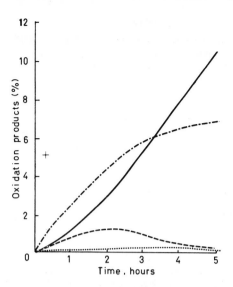

Fig. 18   Yield of oxidation products from p-*xylene as a function of time.*
---- p-*toluic alcohol;* — · — · — p-*toluic aldehyde;* —— p-*toluic acid;* · · · · p-*toluic acid ester.*

molecular weight of the acid chain was increased. The type of solvent also altered the maximum conversion and termination of the reactions.

In the absence of any solvent terephthalic acid begins to form after about 10-25 % of the p-xylene is converted to p-toluic acid. The degree of conversion required depends upon the temperature but in any event the rate of formation of terephthalic acid is 20-25 times slower than that of p-toluic acid. Thus at 40-45 % conversion of p-xylene to p-toluic acid only 1-3 % of the latter was found to be converted to terephthalic acid. It is likely that this low conversion of p-xylene to terephthalic acid in the absence of solvent arises as much from the deactivation of the catalyst by the latter as from the low solubility of p-toluic acid in p-xylene.

The effect of various solvents on the conversion and yield of terephthalic acid from p-xylene using a cobalt stearate catalyst is shown in Table 3. The effect of the nature and concentration of some bromine derivatives on this conversion and yield is shown in Table 4. The data for these tables were obtained at 138°C with 1×10^{-3} moles of cobalt stearate catalyst per litre. The reaction time was 2·5 hours and n-valeric acid was used as the solvent in the systems under study in Table 4.

TABLE 3

Effect of solvent on conversion and yield of terephthalic acid from p-xylene. ($CoSt_2$ conc. 1×10^{-3} mols/lit, $T = 138°C$)

Solvent	Reaction time (h)	Conversion p-xylene (%)	Mol % yield terephthalic acid
nil	2	41·4	1·7
n-butyric acid	2·5	~37	5·2
n-valeric acid	3·5	69·5	~15
isovaleric acid	2·0	~60	7
caprionic acid	2·5	~64·5	8
bromobenzene	2·5	~75	7

TABLE 4

Effect of nature and concentration of bromine derivatives on conversion and yield of terephthalic acid from p-xylene ($CoSt_2$ conc. 10^{-3} moles/lit, $T = 138°C$, $t = 2·5h$, solvent n-valeric acid)

Bromine compound	Concentration (10^{-4} moles/litre)	p-xylene converted (%)	Yield terephthalic acid (%)
nil	—	~58·5	8·7
KBr	3·2	63	~10
NH_4Br	64	71	~12·5
C_6H_5Br	—	65	9
$C_2H_2Br_4$	95·2	~71	11·7
$C_2H_2Br_4$	107·4	~75	~12

Phthalic Anhydride from o-Xylene

The formation of phthalic anhydride from the oxidation of o-xylene in the homogeneous liquid phase is known to occur in accord with the following reactions:

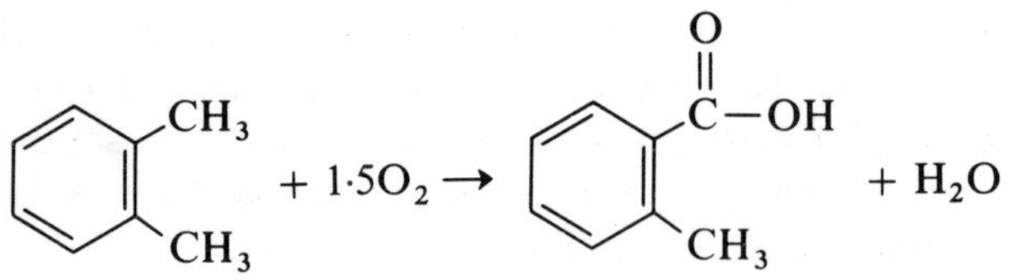

$$2 \quad \underset{CH_3}{\underset{|}{\text{C}}}\!\!-\!OH \quad + CH_3OH \rightarrow \quad \underset{CH_3}{\underset{|}{\text{C}}}\!\!-\!OCH_3 \quad + H_2O$$

$$3 \quad \underset{CH_3}{\underset{|}{\text{C}}}\!\!-\!OCH_3 \quad 1{\cdot}5O_2 \rightarrow \quad \underset{\text{O}}{\text{C}}\!\!\diagdown\!\!\underset{\text{C}}{\diagup}\!O + CH_3OH + H_2O$$

The catalysts utilised for step 1 generally are salts of cobalt, manganese and/or chromium. Catalysts for step 2 may be sulphuric acid or elevated temperatures (250°C) and pressures (85 atmospheres). Those for step 3 generally are cobalt or manganese salts of the C_6-C_{10} fatty acids. At atmospheric pressure the oxidation rate of o-xylene is quite slow. For example a conversion to about 28% o-toluic acid (and 3-4% o-toluic aldehyde) requires 8 hours. At higher pressures (and consequently temperatures) equivalent yields have been attained with reaction times of 30 minutes or less. In relation to the xylene utilised the yield could reach 98%.[153] At a methanol-o-toluic acid ratio of 5 to 1, 250°C and 85 atmospheres pressure, 93-94% esterification was achieved in 20-25 minutes while at 300°C a conversion of 98-100% was attained in only 10-15 minutes. The yield in this case however did not exceed 95%.

The final oxidation (equation 3) is carried out at atmospheric pressure and 150-200°C, with air using small amounts of peroxide initiators to reduce or eliminate the induction period.

Commercially, bromine or various bromides are used as promoters and acetic acid as solvent. The oxidation reactions are carried out at superatmospheric pressures at temperatures between 135 and 275°C.[154]

2.2.3   Non-catalytic Homogeneous Liquid Phase Oxidations

The following are examples of homogeneous liquid phase oxidation reactions which are carried out without the aid of catalysts although in many cases initiators are employed to reduce or eliminate the induction period.

ACETONE FROM ISOPROPANOL

The non-catalytic oxidation of isopropanol to acetone requires somewhat higher temperatures (140 vs 100°C) and pressures (20-21 atmospheres vs 10-12) than the catalytic processes described earlier. Since hydrogen peroxide is formed in the course of the reaction, only small amounts need be introduced initially to initiate the reaction. The yield of acetone ranges between 85 and 90%.

ACETIC ACID FROM PROPYLENE AND n-BUTENE

In the Monsanto Process the propylene is oxidised continuously at 190-210°C at a maximum pressure of 75 atmospheres in accord with the equation

$$CH_3-CH_2=CH_2 + \tfrac{5}{2}O_2 \rightarrow CH_3COOH + CO_2 + H_2O$$

The ratio of propylene to solvent employed ranges between 0·05 and 0·3 (by weight) while the molar ratio of oxygen to propylene in the feed is near 1. The conversion of propylene is about 45% and average yields of about 92% are achieved.[155]

The production of acetic acid from n-butene follows an indirect oxidation route.[156] In the first phase the butene is esterified with acetic acid and in the second the ester is oxidatively split to yield 3 moles of acetic acid per mole of ester. Thus from each mole of butene, 2 moles of acetic acid are formed. The first step is essential to prevent yield losses and contamination of the acetic acid from secondary products formed by butene polymerisation.

The butene used as raw material for this process is obtained by pyrolysis of heavy gasoline fractions. The pyrolysis products are treated to remove ethylene, propylene, isobutene, butadiene and paraffinic hydrocarbons to produce a feed containing about 80% butene.

The butene is esterified over a solid catalyst at 110°C and 15-20 atmospheres pressure. The conversion of butene to ester per pass is 50-80%. The catalyst is removed by centrifugation and following debutanisation the ester and acetic acid mixture is oxidised with air at 195°C and 60 atmospheres pressure. Oxygen enriched air is employed, also. The reaction is strongly exothermic and cooled recycle flows aid in maintaining the requisite temperature.

The reaction products are separated by successive distillations. Intermediate products and non-reacted ester are recycled, the formic acid is generally disposed of by burning. A portion of the pure acetic acid from the top of the column is recycled for temperature control. The main advantage of this process rests in the utilisation of *n*-butene which as a product is secondary to ethylene and propylene. The material requirements for this process are shown in Figure 19.

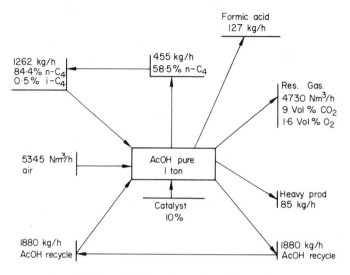

Fig. 19   *Material balance for acetic acid from* n-*butene.*

TRIMELLITIC ACID FROM PSEUDOCUMENE

The manufacture of trimellitic acid by non-catalytic oxidation of 1, 2, 4 trimethyl benzene (pseudocumene) with nitric acid is carried out in four steps.[157] The commercial advantage of this process rests with the fact that pseudocumene is obtained cheaply as a byproduct in the manufacture of *o*- and *p*-xylene.

The oxidation is carried out under pressure in a tabular flow reactor at 180-200°C with dilute nitric acid (about 7%). The nitrogen oxide gases resulting from the reactions are recovered by absorption and along with the diluted nitric acid resulting from the reaction are concentrated for recycle. In the second step the trimellitic acid is separated out by crystallisation, centrifuged, washed, dried and packaged. A third step in the process involves thermic dehydration of the acid to the anhydride at 220°-230°C and the final step is the purification of the anhydride by distillation.

The specific consumptions per ton of trimellitic anhydride are 1 ton pseudocumene, 1·5 tons 100% nitric acid, 100 kWh and 17 tons steam.

Both the anhydride and acid are used in significant and growing amounts in the production of plastics and plasticisers.

MANUFACTURE OF PEROXIDES AND HYDROPEROXIDES

The importance of peroxide and hydroperoxide intermediates in the homolytic production of alcohols, aldehydes, ketones, alkyl oxides and acids has been stated earlier. There is consequently a requirement by the industry for minor amounts of isolated hydroperoxides. These are produced by non-catalytic oxidation in the liquid phase. Some of the hydrooxides isolated from this procedure have been those of methyl cyclohexanol, decaline, methyl cyclohexanone, cyclohexyl benzene, isopropyl naphthalene, isobutyl naphthalene, cumene, ethyl benzene and diisopropyl-benzene.

The technological parameters of a universal pilot plant installation for the manufacture of hydroperoxides must be such that the concentration of the hydroperoxides in the reactor is maintained at a low level at all times. Any accumulation of the hydroperoxide favours the formation of acids which catalyse the decomposition of the hydroperoxides. The procedure adopted is generally as follows. The reaction products from the reactor are naturalized by treating them with a 6% Na_2CO_3 solution in a liquid-liquid extractor. The neutralised product is washed with water to remove Na_2CO_3 and sodium salts of any organic acids and a portion of it is evaporated partly and returned to the reactor. The remainder is treated countercurrently in a liquid-liquid extractor with a polar solvent to extract the hydroperoxide prior to returning it to the reactor. Methanol with 10-20% water is an effective solvent and is preferred because of its low boiling point. In this way an optimum concentration of about 3% hydroperoxide can be maintained in the reactor.[158]

The particulars with respect to the manufacture of the hydroperoxide of methyl cyclohexanol are typical for this process and are listed below.

Reactor capacity	20 litres
Reaction temperature	113°C
Feed rate of (-ol)	10 litres per hour
Feed rate of 10·4% O_2, 89·6% N_2	340 litres per hour
RCOOH in reactor	40 moles per litre
Hydroperoxide in reactor	2·75% in, 3·4% out
Na_2CO_3 solution/water wash rate	10 litres per hour
R-COOH extracted with Na_2CO_3	0·4 grams per hour
CH_3OH feed rate to extractor	2 litres per hour

Hydroperoxide in recycled (-ol)   0·61%
Hydroperoxide production rate   about 38 grams per hour

The composition of the product obtained upon distillation at 31°C and 0·04 mm Hg pressure was:

Hydroperoxide	84·7%
Water	6·8%
Methyl cyclohexanol	0·1%
RCOOH	0·6%
Heptynone and methyl cyclohexanone	7·8%

The preferred operating temperatures for the manufacture of the hydroperoxides of 1-methylcyclohexanone, decaline, cyclohexyl benzene and isopropyl naphthalene are 60, 100-110, 95 and 110°C, respectively. The products obtained have a maximum hydroperoxide content of 90%.

Phenol via Cumene Hydroperoxide

The **BP** Chemicals Ltd process for the manufacture of phenol and acetone from cumene is shown schematically in Figure 20.

According to the disclosed facts the oxidation of cumene is carried out in the reactor 1 at low concentrations of cumene-hydroperoxide because of hydrolytic splitting. The hydroperoxide is concentrated by evaporation

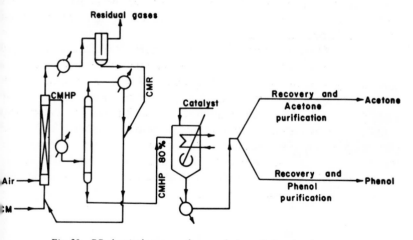

Fig. 20   *BP chemicals process for manufacture of phenol and acetone.*
1—cumene oxidation reactor; 2—cumene hydroperoxide recovery unit; 3—cumene hydroperoxide hydrolysis unit; 4—gas liquid separator; CM—cumene; CMHP—cumene hydroperoxide; CMR—cumene recycle.

and hydrolysed with an acid catalyst in vessel 3. Both the cumene recovered from the residual gases and that from the evaporator are recycled to the base of the reactor. Alpha-methyl styrene which is formed as a secondary product may be isolated and recovered or hydrogenated to cumene and recycled. The yield of phenol based on cumene is about 92·5% and the weight ratio of acetone to phenol in the product is about 0·6 to 0·66 to 1.[159]

Another process of hydroperoxidation which shows commercial promise is the simultaneous manufacture of styrene and propylene oxide by the following chemistry.[160]

ethyl benzene $\rightarrow$ ethyl benzene hydroperoxide

propylene + ethyl benzene hydroperoxide $\rightarrow$

propylene oxide + methyl benzylic alcohol

methyl benzylic alcohol $\rightarrow$ styrene

The reaction between propylene and ethyl benzene hydroperoxide occurs best in the presence of certain selective catalysts which give high conversions and yields. For example, the respective conversions achieved in 60 minutes at 110°C with naphthanates of titanium, tungsten and molybdenum were 54, 80 and 95%. The nature of the acid component also influences the conversions. That achieved (under the above conditions) with phospho-molybdic acid was 67% and with lithium phospho-molybdenate only 45%.

The type of hydroperoxide present also influences the conversion; that of ethyl benzene gave a 92% conversion while that of cumene gave 79%. The hydroperoxidation of ethyl benzene at 130°C and 3·5 atmospheres occurs with a selectivity of about 84%. The epoxidation of propylene at 100-130°C and 1-3 hours reaction time occurs with about 71% selectivity. Finally the dehydration reaction of methyl benzylol at 200-250°C in the presence of a titanium based catalyst is about 80% complete. Acetophenone formed as a co-product in this step is separated out along with the unreacted methyl benzylol and following hydrogenations (to the alcohol) is recycled to the dehydrogenator.

CHAPTER 3

Heterogeneous Catalytic Oxidation of Aromatic Hydrocarbons, Alkyl Benzenes and Olefins

3.1 MANUFACTURE OF PHTHALIC ANHYDRIDE

3.1.1 Introduction

Phthalic anhydride has many multiple uses. In 1965 from a total production of 222 000 tons, the usage was about 79 000 for resins, 87 000 for plasticisers, 37 000 for polyesters and 9000 for dyes and other minor uses. For many years the naphthalene obtained from coal coking operations (about 25 kilograms naphthalene per ton of coke) was used as the raw material. With the rise of the petrochemical industry both o-xylene and naphthalene from gasoline dealkylation processes have become competitive raw materials on a world wide basis.

The oxidation of o-xylene or naphthalene can be carried out in the gas phase over solid fixed or fluidised catalysts. The kinetics of the oxidation reactions of o-xylene in the liquid phase have been treated in the previous chapter. In Tables 5 and 6 the main, secondary and potential products of oxidation of naphthalene and o-xylene are listed along with the number of moles of oxygen required for their formation and the heats of reaction per gram atom of oxygen consumed.[161] It should be noted that the overall oxygen requirement and hence the total heat of reaction is appreciably greater for naphthalene as a raw material than for xylene. The heat released per gram atom of oxygen, however, is greater for the o-xylene oxidations and is greatest for the initial intermediate stages. In the case of naphthalene oxidation the heat release per gram atom of oxygen is lower for the initial or intermediate stages of oxidation.

3.1.2 Technological Kinetic Considerations

The following conclusions appear to be consistent with data and observations reported in the literature on the kinetics and production of phthalic anhydride by heterogeneous catalytic oxidation of naphthalene or o-xylene.[162-187]

TABLE 5
Products, oxygen requirements and heats of reaction in the oxidation of naphthalene

Main products	Secondary products	Potential products	Oxygen requirement g atom/mole	Heat of reaction kcal/g atom O_2
	1 naphthol		1	47·1
1,2-naphthoquinone			3	42·0
1,4-naphthoquinone			3	43·9
		o-carboxy-allocinnamic acid	4	48·9
		phthalonic acid	8	50·7
phthalic anhydride			9	49·9
	phthalic acid		9	51·3
	benzoic acid		9	51·3
maleic anhydride (2 moles)			12	41·2
	acrylic acid (2 moles)			48·1
	ethylene (2 moles)		12	49·1
maleic anhydride			18	49·9
	acrylic acid		18	50·3
	ethylene		18	50·6
carbon dioxide, water			24	51·3
			Average	48·4

The kinetics of the oxidation of naphthalene can be expressed by an equation of the form[163,164]

$$-\frac{dP_{Hc}}{dt} = \frac{1}{K_1 P_{Hc} + \dfrac{\beta}{K_2 P_{O_2}^n}} \tag{1}$$

where

K_1 = rate constant for disappearance of hydrocarbon

K_2 = rate constant for disappearance of oxygen

P_{Hc} = partial pressure of hydrocarbon (mm of Hg)

P_{O_2} = partial pressure of oxygen (mm of Hg)

β = stoichiometric ratio of O_2/hydrocarbon

TABLE 6
Products, oxygen requirements and heats of reaction in the oxidation of O-xylene

Main products	Secondary products	Potential products	Oxygen requirement g atom/mole	Heat of reaction kcal/g atom O_2
		o-tolubenzyl alcohol	1	56·7
o-tolualdehyde			2	54·5
		o-toluic acid	3	54·3
		o-hydroxy methyl benzoic acid	4	51·1
phthalic anhydride			6	51·4
		phthalic acid	6	53·5
		benzoic acid	6	53·4
2 moles maleic anhydride			9	47·1
	2 moles acrylic acid		9	48·5
	2 moles ethylene		9	49·8
maleic anhydride			15	50·5
	acrylic acid		15	50·9
	ethylene		15	51·3
carbon dioxide water			21	52·0
			Average	51·8

At constant P_{O_2} equation 1 reduces to

$$\frac{dp_{Hc}}{dt} = \frac{K_1 P_{HC}}{1 + K_1 C P_{HC}} \qquad (2)$$

where $C = \beta/K_2 P_{O_2}$ in which β represents the number of molecules of oxygen required per mole of hydrocarbon and K_2 is in effect the re-oxidation constant of the catalyst with oxygen.

From equation 2 it can be seen that at small values of P_{HC} the rate of reaction is first order with regard to the hydrocarbon and at higher values tends to zero. Thus at a partial pressure of 0·4 mm Hg of naphthalene in the feed over a catalyst comprised of 9% V_2O_5, 21% K_2SO_4 and 80% SiO_2, the reaction order was found to be 1 whereas at 8 mm Hg partial pressure of naphthalene the reaction order found was zero.[163] In commercial operation where the partial pressure of hydrocarbon was varied between 15 and 76 mm Hg, a zero order rate of reaction with respect to the hydrocarbon has been noted, also.[162]

The rate of reaction at 337°C with a partial pressure of naphthalene at 10-16 mm Hg pressure was found to have the value 1.71×10^{-4} moles per gram catalyst per hour. At low concentrations of naphthalene the rate becomes first order with respect to oxygen and is given by $K_2 P_{O_2}$. As the oxygen concentration is increased the amount of the intermediate product naphthoquinone relative to the final product phthalic anhydride is decreased.

The addition of K_2SO_4 to the V_2O_5 catalyst has been shown to decrease its absorption capacity for naphthalene and oxygen, to increase the rate of exchange and decrease the energy of activation.[168,169]

The oxidation of o-xylene over V_2O_5 catalyst proceeds in accord with the following successive reactions[170]

Reactions 1 and 2 are first order with respect to o-xylene and o-toluic aldehyde respectively. Reactions 3, 4 and 5 are zero order with respect to the hydrocarbon but like 1 and 2 are proportional to the square root of the oxygen concentration.

The types of catalysts employed for the production of phthalic anhydride are generally based on V_2O_5 with additions of K_2SO_4. Silica gel supports are most common probably due to their low porosity which prevents undue secondary reactions of complete oxidation. Characteristics of some typical catalysts used in naphthalene oxidation are listed in Table 7.[171]

TABLE 7
Characteristics of catalysts used for naphthalene oxidation

Basic composition	Operating temperatures (°C)	Contact time (Sec.)	Observations, yield, etc.
40-70% SiO_2 50-20% $K_2S_2O_7$ 3-25% V_2O_5	350-400	3-20	fluidised bed operation (f.b.o.)
$SiO_2 + V_2O_5$	290	0·3	90% phthalic anhydride air/naphthalene 3X theory
V_2O_5	—	—	85% phthalic anhydride
V_2O_5 microspheres	—	—	84-87% phthalic anhydride (f.b.o.)
V_2O_5 fused	565	—	87% phthalic 10% maleic (f.b.o.)
$V_2O_5 + SiO_2$	540-550	0·25	79-82% phthalic anhydride
20% V_2O_5 on SiO_2	400-550	0·05-0·5	50-80% phthalic anhydride (f.b.o.)
$TiO(VO_3)_2$ with V:Ti = 1·5:3	—	—	85-90% phthalic anhydride
Phospho vanado-ammonium tungstate	430-470	0·5	80% conv. of naphthalene

The technology of catalyst preparation is an intricate art. Fused catalysts or catalyst 'fusions' which incorporate melted V_2O_5, dispersed V_2O_5 or dispersed mixtures of V_2O_5 and certain promoters are obtained generally by heating ammonium metavanadate (and other additives) with an inert support such as alumina granules or silicon carbide. Silica gel supports for dispersed catalysts are prepared most commonly by fusing Kieselgur or SiO_2 with potassium hydroxide at about 950°C using a molar ratio of SiO_2 to KOH between 3·8 and 4 to 1. This fused mixture is then treated with a 25 % solution of ammonia and the resulting solution of potassium silicate is neutralised with 16° Bé H_2SO_4. The moist gel thus obtained is ground and dispersed with enough 25 % NH_3 solution to produce a slurry with a pH of 8 before being dried at 120°C.

Catalysts which contain promoters are of the vanadate or metal oxide type.[172] Mixed oxide catalysts can be obtained by decomposition of mixed metal oxalates at 400-440°C. For example suitable mixed metal oxide catalysts have been obtained by decomposing the oxalate salts obtained from oxalic acid solutions of ammonium metavanadate, to which ammonium molybdate or cobalt, uranium or cerium nitrate was added. Vanadate promoters are obtained also by treating ammonium meta-vanadate solutions with stannic chloride or silver nitrate to produce precipitates on the catalytic surface.

Irrespective of the nature of the catalyst and its support the technological requirement remains the same. The catalyst must be highly active, highly selective and highly stable under mechanical or thermal stress and with respect to poisoning by trace materials.

The BASF Co. has evaluated catalysts to meet the above requirements for the oxidation of o-xylene. They note that when the V_2O_5 which comprises the main component of the catalyst is present in finely divided form, advanced or total oxidation reactions are favoured. One way of obviating this was to reduce the contact time, another was to add alkali sulphates to the catalyst. However in the latter case, although yields of phthalic anhydride as high as 97 % were obtained, the corrosion of the reactor system by SO_3 (which is formed at temperatures as low as 360-380°C) was appreciable. Moreover because of the lower activity of this type of catalyst larger reactor vessels are required.

A catalyst with high activity and good selectivity for fluidised and fixed bed operations consists of mixed oxides of TiO_2 and V_2O_5.[173] The optimum proportion of the V_2O_5 in the active mass was found to vary between 2 and 15 % for thicknesses of the active layer on the support between 0·03 and 1·5 mm and a maximum ratio of V_2O_5 to total catalyst mass of 0·03. Because of this low ratio of V_2O_5 to the bulk of the catalyst a typical reactor installation ($\sim 15\,000$ tons product/yr.) containing 10 000 tubular elements requires less than 100 kg of V_2O_5. Again because of the high activity of this catalyst it has been used to coat smooth surfaced large spheres (6-12 mm in diameter) to produce highly active fixed bed catalysts with low fluid flow resistances. As an example of the effect of catalyst pellet size on the flow resistance, a pressure drop of 540 mm of water column across a bed of 5 mm spheres at a given flow rate of air and ydrocarbon vapour through it, will be reduced to 74 mm when 9 mm diameter spheres are substituted.

Optimal activity of this catalyst is achieved by heating the coated spheres at a temperature no higher than 200°C at the time of preparation. Additions of small amounts of lithium, zirconium and phosphorus compounds to the bulk of the catalyst enhance the yield and purity of the phthalic anhydride formed.

The recent research studies on catalysts used for oxidation of naphthalene to phthalic anhydride under fluidised bed conditions are of interest in this context.[174] The catalysts for these studies were prepared by grinding and sieving silica gel with large pores into fractions ranging from 0·08 to 0·43 mm, impregnating these fractions with various aqueous salt solutions at 60°C, aging these for 3-4 hours at 20-30°C, drying the suspensions by gradually raising the temperature to 130°C over a 10 hour period, and finally baking this material for 50-60 h at 330-400°C after regrinding to the desired size. The salt mixtures chosen were as follows:

(a) $VOSO_4 + AgNO_3 + K_2SO_4$

(b) $VOSO_4 + (NH_4)_6Mo_3O_{23}.4H_2O$

(c) $VOSO_4 + KBr + K_2SO_4$

(d) $VOSO_4 + H_3PO_4 + K_2SO_4$

(e) $VOSO_4 + K_2SO_4 + KHSO_4$

These studies showed that catalysts prepared with solution (a) had good selectivity but low activity. Those from solution (b) had somewhat limited selectivity. Although those from solution (c) were most selective the optimum performance was obtained from the catalysts prepared from solutions (d) and (e).

The catalysts prepared from the latter solution contain potassium pyrosulphate and no doubt owe their superior performance to this compound. As the combined content of K_2SO_4 and $K_2S_2O_7$ was increased to about 26 % for a $K_2O:SO_3$ ratio between 1 and 1·7 it was found that the selectivity and resistance to abrasion of these catalysts increased while the effective working temperature decreased to 330-340°C or lower. For example, a 93 % conversion of naphthalene to phthalic anhydride with a conversion to maleic anhydride of less than 0·9 % was obtained at 310-320°C when mixtures of naphthalene and air (in ratios from 1:10 to 1:15) were fed at specific loadings of 40 grams naphthalene per kilogram catalyst per hour and a contact time of 12 seconds.

3.1.3 Technological Aspects of Naphthalene and *o*-Xylene Oxidations in Fixed and Fluidised Bed Reactors

FIXED BED LOW TEMPERATURE OXIDATION PROCESSES

The oxidation of naphthalene to phthalic anhydride in fixed bed reactors is generally carried out at about 360°C when the catalyst is fresh and up to about 385°C as the catalyst ages.[176-179] Multi-tubular reactors with as many as 3000-10 000 tubes per reactor are employed to effect high catalyst surfaces and high heat exchange surfaces. A typical tube might be 25 mm in diameter and contain about 1·2 litres of catalyst. A typical fixed bed flow process is shown in Figure 21.

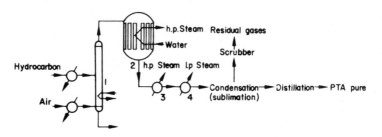

Fig. 21   *General flow diagram for manufacture of phthalic anhydride by fixed bed catalytic 1—evaporator and reactant premixer; 2—fixed bed catalytic reactor; 3 and 4—heat exchangers; Hydrocarbon—generally naphthalene but may be o-xylene; PTA—phthalic anhydride.*

For this process naphthalene of high purity is required. As can be seen from Figure 1, the naphthalene concentration in the air stream to the reactor is regulated by altering the flow rate of these two components since the evaporator 1 surface is large enough to permit total evaporation up to the capacity of the reactor. The air and naphthalene vapour mixture enters the reactor at 140-150°C. Because the explosive limit of naphthalene-air mixtures at the reactor temperature ($\sim$370°C) is about 50 grams naphthalene per cubic metre of air, the concentration of the feed stream usually is maintained at about 37 grams naphthalene/Nm^3 air ($\equiv$ 1 kg:35 kg). The volumetric space velocity is generally about $1000\,h^{-1}$ for a contact time of 1·33 seconds.

Of the heat released by the reaction about 50% is carried off by the gas vapour products, 20% is lost by radiation and the remainder is removed by heat exchangers 3 and 4 producing high and low pressure steam. The heat exchange fluid in the reactor is usually a fused salt (e.g.

potassium nitrate-sodium nitrite) although mercury and high boiling organic liquids (diphenyl oxide) also have been employed. The temperature differential between the reaction zones and the bath does not exceed 20-40°C unless an operating problem arises.

In the case of o-xylene oxidation to phthalic anhydride the feed concentration is 40-65 grams of xylene per Nm^3 of air. The inlet temperature of the reactants is the same as for naphthalene feed (~ 150°C), but the reaction zone temperatures span a somewhat broader range (350-420°C). Injection of small quantities of SO_2 into the feed stream has been reported to reduce this large variation in the reaction temperature.[161,178] For a typical catalyst (same as for naphthalene) with a contact time of 0·15 seconds, the average yield of phthalic anhydride is about 60%, of maleic anhydride 7·5%, aromatic aldehydes 3·4%, CO_2 about 20% and CO $\sim$ 7·5%. Minor amounts of various other intermediate oxidation products account for the remainder.

FIXED BED OXIDATION PROCESS AT HIGH TEMPERATURES

The high temperature oxidation processes differ from those described above primarily in that (a) naphthalene of low purity may be used, (b) the reaction temperatures used range between 420 and 550°C; (c) small size reactors are employed since the volumetric space velocities range between 4000 and 5000 h^{-1}, and contact times of the order of 0·6 seconds are employed; (d) catalysts of lower activity are effective (because of the higher reaction temperatures). A catalyst comprised of 7-8% V_2O_5 on an inert support is typical;[180] (e) the feed concentration of naphthalene may be as high as 63 grams per Nm^3 of air which is near the explosive limit of this mixture at operating temperatures. The reactors are equipped therefore with explosion discs.

Although boiling mercury at 2 to 8 atmospheres pressure the preferred reactor cooling system for this process in earlier installations is still used, most recent installations employ fused baths of potassium nitrate and sodium nitrite because of the cost and toxicity of mercury and because fused salt reactors are simpler to design and construct.

FLUIDISED BED OXIDATION PROCESSES

The oxidation processes in a fluidised bed differ significantly in a number of ways from that in fixed bed operations. The particular advantageous characteristics of the fluidised bed reactor lie in its ability to transfer heat at a high rate and to damp out explosive waves. Thus up to 115 grams of naphthalene per Nm^3 of air may be fed to the reactor without danger of explosion and concern that the temperature difference within the reaction zone will be greater than 2-6°C. In the case of o-xylene oxidation the process is carried out in the fluidised bed reactor with 50 % less excess air than is used in the fixed bed process. Thus, with both raw material sources the effluent from the fluidised bed reactor is richer in phthalic anhydride resulting in a considerable simplification of product recovery and purification.

A typical catalyst for the fluidised bed process may consist of 10 % V_2O_5, 20-30 % K_2SO_4 and 60-70 % silica gel. The granular dimensions range between 50 and 200 μ.[181]

Other characteristics of fluidised bed reactors have been described in the literature.[182] A high productivity reactor is characterised by a catalyst circulating system outside the reactor and a heat exchanger at the lower part of the reactor where the major part of the heat of reaction is released. The velocity of the gas stream within the reaction zone is about 6 m/sec. The productivity of phthalic anhydride is generally of the order of 125 $kg/m^2/h$.

A schematic flow diagram for phthalic anhydride manufacture from naphthalene in a fluidised bed reactor system is shown in Figure 22.

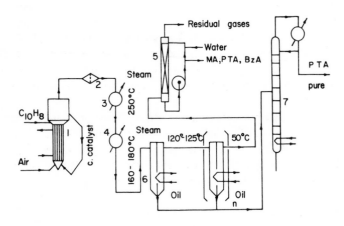

Fig. 22   General flow diagram for manufacture of phthalic anhydride by fluidised bed catalytic oxidation of naphthalene.

1—catalytic fluidised bed reactor; 2—filter; 3 and 4—heat exchangers; 5—scrubber; 6—condenser-sublimators; 7—distillation column; MA—maleic anhydride; PTA—phthalic anhydride; BzA—benzoic acid.

The following comments on this process will clarify its pertinent features and operations. The product stream from the reactor is passed through a filter 2 and through heat exchangers 3 and 4 where high and low pressure steam are produced. The gases and vapours exiting from the final heat exchanger 4 enter at 160-180°C a series of condenser-sublimators 6 where the phthalic anhydride is first deposited in solid state and then sublimed to a distillation-rectification column 7. Oil is utilised for cooling and heating the condenser-sublimators.

The gaseous stream from the condenser-sublimators is composed largely of nitrogen, carbon oxides and water, but also entrains vapours of phthalic anhydride, maleic anhydride, benzoic acid and minor amounts of other intermediary oxidation products. The vapours are recovered by a scrubber 5.

In operating the condensers, it is important to ensure that the temperature does not drop below 40°C. At temperatures below this maleic anhydride and naphthoquinone condense out, thereby affecting the purity of the phthalic anhydride. Moreover, at temperatures below 40°C, the deposits become wet and corrosion is increased.

The product from the sublimator is about 98·5 % phthalic anhydride, 0·25 maleic anhydride and 0·25 % benzoic acid. The first distillation of the sublimate is carried out at about 200°C and 40-80 mm Hg pressure. When a second column is employed the distillation is carried out at somewhat higher temperature ($\sim$220°C) and pressure ($\sim$150 mm Hg). The final purity of phthalic anhydride is at best 99·7 %.

When impure naphthalene is used in the feed stream the phthalic anhydride may be purified with 98 % H_2SO_4. The acid requirement is of the order of 0·5-1 % of the product.[183] Residues of phthalic anhydride accumulated at the base of the distillation columns may be recovered by distillation at high vacuum or by atomisation of the fused material with a stream of N_2 or CO_2 gas.[184] The energy requirement is 60-65 kilocalories per kilogram of phthalic anhydride when the residue contains about 70 % phthalic anhydride. The main operating conditions for the fixed bed, fluidised bed and homogeneous liquid phase processes for the manufacture of phthalic anhydride are summarised in Table 8.

The data in this table indicate quite clearly that the homogeneous liquid phase oxidation process to produce phthalic anhydride from o-xylene is not competitive with either the fixed bed or fluidised bed processes. This is even more strongly evidenced when it is noted that the liquid phase process is exclusive for o-xylene while the other two processes also may utilise naphthalene as raw material.

TABLE 8
Comparison of reaction conditions and yields of phthalic anhydride manufacturing
processes

Reaction conditions	Naphthalene fixed bed		o-Xylene		
	Low T	High T	Fixed bed	Fluidised bed	Liquid-phase
Temperature (°C)	360-385	420-550	350-550	370	120-275
Pressure (Atmos.)	0·5	2·0	1·75	1·05	10-40
Contact time (sec.)	1·33-1·5	0·5-0·65	0·15	0·5-3	190
HC/Air (g/Nm³)	37-43	50-63	40-65	85-95	—
Load (kg/m³hr)	34	320	250	16	—
Cat./H.C. (kg/kg)	13·4	6	8	37	—
Phthalic Anhydride Yield (%)	86-90	65-75	50-65	69-75	68-72

The fluidised bed process allows for a higher yield, offers less risk of explosion and also permits the best recovery of heat.[188] Its main disadvantages lie in the fact that the catalyst erodes the cooling element within the reactor and is lost in significant amounts due to attrition.

Irrespective of the process, however, the use of o-xylene as the feed material *in lieu* of naphthalene offers several distinct advantages. First of all it simplifies the problem of feed composition and flow control. Secondly since the exothermic heat of reaction is about 16% lower, the reactor design and operation is easier. Thirdly the lower air requirement permits smaller unit installations with attendant reduction in captial and operating costs. Finally 1:4 naphthoquinone cannot be formed as a secondary product from o-xylene thereby permitting a simpler and more effective purification of the phthalic anhydride.

Three processes are carried out on a worldwide basis at present.[185] The Sherwin Williams-Badger Co. Inc. process utilises high purity naphthalene and effects the oxidation in a fluidised bed reactor at 340-385°C. The yield by weight is about 0·98 parts phthalic anhydride per part reactant feed. The United Coke Chemicals Ltd-Foster Wheeler Corp. process also utilises naphthalene as a raw material and fluidised bed technology. The BASF-Foster Wheeler Corp. process on the other hand employs o-xylene as raw material and fixed bed reactor technology. The yield in this case by weight, is about 1 part phthalic anhydride per part of o-xylene feed.

3.2 ETHYLENE OXIDE

3.2.1 General Information

Ethylene oxide is an intermediary chemical with multiple uses. Of the 1 350 000 tons per year which are produced today on a worldwide basis, 50-60% is used in the manufacture of ethylene glycol. The remainder is converted to surface active agents, polyethers, di, and triethanolamines as well as diverse mono- and diethylamines with aromatic substituents.

About 40% of the world production is manufactured today by direct catalytic oxidation of ethylene with air or oxygen. The primary and secondary reactions are:

$$CH_2{=}CH_2 + \tfrac{1}{2}O_2 \xrightarrow{Ag_2O_2} CH_2\!\!\overset{\displaystyle O}{\overset{\diagup\ \diagdown}{-}}\!\!CH_2 + 21\ kcal/mol$$

$$CH_2{=}CH_2 + 3O_2 \longrightarrow 2CO_2 + 316\ kcal/mole$$

Other processes for the manufacture of ethylene oxide on a commercial scale include hydrolysis of ethylene chlorohydrin with $Cu(OH)_2$[204,205] hydroperoxidation of ethylene[206,208] and catalytic oxidation of ethanol with air. These processes are of lesser importance and will not be treated in this text.

The ethylene epoxidation is unique in that, to date, the only catalysts which have been found to have adequate activity and selectivity for this reaction are silver based. Moreover their selectivity is limited to ethylene. Other olefins when treated on these catalysts under similar conditions are oxidised totally. In view of this a brief look at the known catalyst characteristics is in order at this time.

3.2.2 Catalyst Characteristics

As stated earlier, silver based catalysts are unique for the epoxidation of ethylene, and ethylene is unique to the epoxidation reaction. The usual range of reactor conditions is 200-300°C and 1-20 atmospheres pressure.

The form of catalyst employed is determined primarily by the reactor system, although powder suspensions and supported catalysts may be employed in fixed and fluidised bed reactors. The activity of a catalyst system is difficult to determine because of carbon dioxide release, and special installations have been designed to determine the activity of several samples simultaneously at 1-10 atmospheres under isothermal continuous flow conditions.[228] The oxides and carbonates of barium, magnesium and copper are frequently used as promoters. The peroxides of the alkaline earth metals and oxides of gold, iron and manganese also may be used for this purpose. Further pertinent particulars are summarised in Table 9.

TABLE 9
Characteristics of epoxidation catalysts

Oxidation agent	Catalyst form	Reaction temp. (°C)	Conv. (%)	Yield (%)	Ref.
Air or O_2	$Ag_2O-BaCO_3.MgCO_3$	237	—	61-68	190
Air	Ag on Al_2O_3 or SiC	270-276	90-97	54-59	191
Air	$Ag \sim Na_2CO_3.CO(Ac)_2$ $13.4:17:1:40.4$ firebrick	210	31	46	192
Air	Ag metal	200-300	—	70	193
Air	$Ag_2CO_3-CuCO_3$ on support	210	45	65	194
O_2	Ag_2O-CuO on Silica gel	—	40	90.3	195
Air	Ag on Al_2O_3	270	—	75	196
O_2	$Al_2O_3-V_2O_5$	400-600	90-100	—	197
Air	Ag_2O-CeO_2 on support	330	77	—	198
O_2 and inert gas	Ag_2O-CaO or BaO	250	60	—	199
O_2 and inert gas	35% Ag on graphite	284	54	68	200
Air	95% Ag 5% Silica gel	240	48	64	201
Air	Ag	205	44	50	202
O_2	Ag	220-350	23-85	—	203

3.2.3   Reaction Kinetics of Ethylene Epoxidation

The oxidation kinetics of ethylene have been studied by numerous researchers. The evidence from these studies indicates that the epoxidation of ethylene occurs simultaneously with other oxidation reactions which lead ultimately to the formation of carbon dioxide and water. The reactions are essentially irreversible because the free energies of formation of ethylene oxide and carbon dioxide and water from both ethylene and ethylene oxide have large negative values at 25 and 200°C.[209] It is difficult to establish an accurate kinetic expression because certain reactions are inhibited not only by catalytic impurities, but also by carbon dioxide, water and ethylene oxide.

The rates of formation of ethylene oxide and of carbon dioxide are proportional to the concentrations of ethylene and oxygen in both fixed bed and fluidised bed reactors.[210-212]

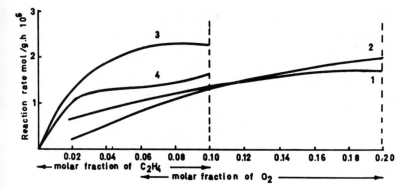

Fig. 23   Effect of ethylene and oxygen concentration on the rate of formation of ethylene oxide
and carbon dioxide in a fixed bed reactor.
Graphs 2 and 4—rate of formation of ethylene oxide. Graphs 1 and 3—rate of formation of
carbon dioxide.

The rates of formation of ethylene oxide and of carbon dioxide as a
function of ethylene and oxygen concentration are graphed in Figure 23
for the fixed-bed reactor, and Figure 24 for the fluidised bed reactor.

The conversion of ethylene has been found to increase as the temperature
is increased. The yield of ethylene oxide also increases initially but quickly
reaches a maximum. For example at about 255°C a conversion of 28%
and a yield of 45% were found whereas at about 285° where the conversion
was maximum the yield had dropped off to 40%, and with further increase
of temperature to 300-310°C the yield fell to 28-30%.[213] The foregoing
results were obtained using a molar ratio of 10:1 for air:ethylene and a

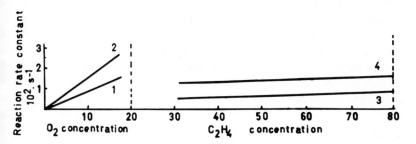

Fig. 24   Effect of ethylene and oxygen concentration on the reaction rate constants for
formation of ethylene oxide and carbon dioxide in a fluidised bed reactor.
Graphs 2 and 4—rate constants for ethylene oxide formation; graphs 1 and 3—rate constants
for carbon dioxide formation.

contact time of 1 second. Above 285°C the conversion also was found to decrease presumably due to the inhibiting effects of water, carbon dioxide and ethylene oxide.

In practice, reactors are sized on the basis of empirical kinetic equations deduced from specific experimental data. The following expression for rate of ethylene oxide formation over the catalyst Ag_2O on α-Al_2O_3 is typical[214]

$$\frac{dn}{dt} = kP_{C_2H_4}^a P_{O_2}^b$$

where

n = moles of ethylene oxide formed per gram of catalyst

t = time in hours

p = partial pressure of reactants as shown

and the following empirical values for a, b and k apply:

	240°C	260°C
a	0·365	0·316
b	0·667	0·667
k	0·0071	0·0141

3.2.4 Technological Aspects and Comparisons of Ethylene Oxide Manufacture

The oxidation of ethylene to ethylene oxide can be effected, as shown in Table 9, with air, oxygen enriched air, oxygen and inert gas, or pure oxygen. The amount of purging required to maintain a constant CO_2 content in the recycle stream will be determined by the oxygen content of the oxidising agent used.[229] As illustrated in the flow diagram of Figure 25, ethylene and oxygen are premixed with recycle gas in a mixer 1 prior to introduction into a reactor 2. The effluent from the reactor is separated in an absorber-scrubber into a liquid fraction of impure epoxide and a gas stream. The latter is purged and recycled whereas the former is sent to a recovery-purification section. When air is used the amount of purging required is about ten times that when oxygen is employed, consequently ethylene oxides yields are lower with air than with oxygen.[215]

Small additions of organic halides (dichloroethane) into the feed stream have been found to increase appreciably the yield of ethylene oxide.

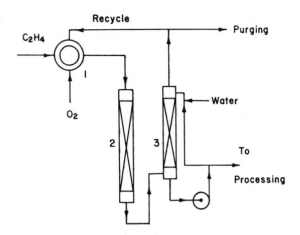

Fig. 25 General flow diagram for manufacture of ethylene oxide by catalytic oxidation of ethylene.
1—reactant premixer; 2—catalytic reactor; 3—scrubber; (O₂ source may be air; oxygen enriched air or pure oxygen).

For example at 283°C with an air to ethylene molar ratio of 10·5 to 1 and a contact time of 0·9 seconds the addition of 1 % dichloroethane increased the conversion relative to ethylene oxide by about 25-28 %. However this effect was totally cancelled when halogen content was further increased about threefold.[216] It is probable that the halogen compounds inhibit the formation of carbon dioxide. Other agents to effect this inhibition also have been proposed. Among these are polytrifluoroethylene[217] chlorinated aromatic compounds[218] halogen compounds of potassium and sodium[219] cyanides of metals from group 1 and 8[220] and compounds of sulphur, selenium and tellurium[221] which are incorporated into the catalyst.

Optimum yields are obtained at reduced temperatures (220-240°C) with 3-5 % ethylene and 5-10 % oxygen in the gas feed. Under these conditions the conversion ranges between 28 and 40 % and the selectivity between 60 and 70 %. The concentration of ethylene oxide in the reactor gases varies between 0·8 and 1·5 %.

The main problem in industrial reactor design is to allow for efficient removal of the large amount of heat which is released by this reaction (about 55 kcal/mole). At a normal productivity level of 0·2 kg ethylene oxide per litre of catalyst per hour and a selectivity of about 60 %, approximately 700 kcal per litre of catalyst per hour must be removed to maintain the reactor temperature within acceptable limits. In fixed bed reactors this is generally achieved by employing multitubular reactors which contain between 3000-3500 catalyst-filled tubes with diameters as small as 2 cm.

The manufacture of ethylene oxide by catalytic oxidation of ethylene with air is generally carried out at about 250°C and 15 atmospheric pressure. In general the productivity is increased by carrying out the oxidation in two steps[222] as shown in Figure 26.

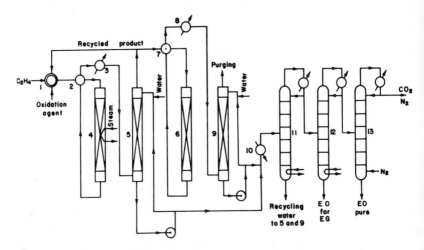

Fig. 26   Extended general flow diagram for manufacture of ethylene oxide by catalytic oxidation of ethylene in two steps.
1—reactant premixer; 2,3,7,8 and 10—heat exchangers; 4—primary catalytic reactor; 5 and 9—scrubbers; 6—secondary catalytic reactor; 11 and 13—desorption columns; 12—distillation column (oxidation agent—air in this case).

In this case ethylene and air at about 15 atmospheres pressure are premixed with recycle gases which contain about 80 % N_2 and 9-10 % O_2 and preheated to about 145°C by heat exchanger 2, then fed to the main reactor 4. The ethylene content of the feed at this stage is about 4·5 % and the oxygen content 7-8 %. The reactor gases which exit at about 250°C must be precooled by exchangers 2 and 3 to about 20°C to permit effective absorption of the ethylene oxide by scrubber 5, which operates at about 12 atmospheres pressure.

The portion of the gases from the scrubber which is not recycled to the feed is passed to the secondary reactor 6 where the second oxidation step is carried out. The conversion in the first reactor is usually limited to about

23 % in which case a yield of about 65 % is obtained. In the second reactor conversions generally approach 40% and yields decline to about 55%.

The water solutions from the scrubbers 5 and 9 contain about 1% ethylene oxide. They are combined, preheated by heat exchanger 10 and sent to the absorbing column 11 where the ethylene oxide is desorbed at about 80°C and 0·4 atmospheric pressure. The vapours from column 11 are compressed and redistilled under pressure in column 12. The ethylene oxide drawn from the base of the latter column is 90-92% pure and is generally used for ethylene glycol manufacture. The ethylene oxide vapours emitted from column 12 are condensed at $-11°C$ and separated from CO_2 in column 13 by desorption with nitrogen gas at $-8°C$ and 1·3 atmospheres.

The manufacturing of ethylene oxide via ethylene chlorohydrin compares favourably with that by direct catalytic oxidation on some factors and unfavourably on others. For example C_2H_4 requirements are 0·75-0·8 tons per ton of ethylene oxide for the chlorohydrin route and 1 ton per ton for the direct catalytic oxidation. However the former process also requires up to 2 tons of calcium oxide per ton of epoxide. With respect to energy requirements the direct oxidation process has a distinct advantage. The electrical energy and steam requirements per ton of ethylene oxide respectively are 1750 kWh and 0·6 tons for the direct oxidation process versus 200 kWh and 10 tons for the indirect route. Clearly these and other factors must be evaluated and compared to establish the real techno-economic merits of each process at a given plant location.

The main processes employed internationally for the manufacture of ethylene oxide are those of the Scientific Design Co. Inc., Nipon Shokubai Kagaku Kogyo Co. Ltd, Shell Development Co. and Chemische Werke Huels A.G.

The Scientific Design Co. process flow diagram is shown in Figure 27. In this process the gases from the reactor 1 are cooled by the recycle gases from the scrubber 2 where ethylene oxide is dissolved from the cooled reactor effluent stream under pressure. A part of the recycle is treated to remove CO_2 prior to reuniting it with the remainder for recycle to the reactor. The latter and the silver catalyst employed are designed to utilise either air or oxygen as the oxidising agent.[225]

The water solution of ethylene oxide is preheated by water taken from the base of the desorber 5 where the ethylene oxide is desorbed, then cooled in the stripper 6 and distilled to produce a high purity product in column 7.

The Nipon Shokubai Kagaku Kogyo Co. Ltd process is similar to the above process but includes the manufacture of mono-, di- and triethylene

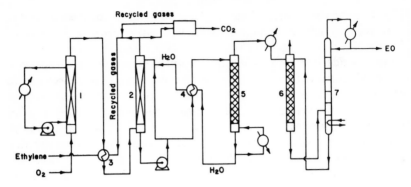

Fig. 27 Scientific Design Co. process for manufacture of ethylene oxide.
1—catalytic reactor; 2—scrubber; 3 and 4—heat exchangers; 5—desorption column; 6—stripping column; 7—distillation column; EO—ethylene oxide.

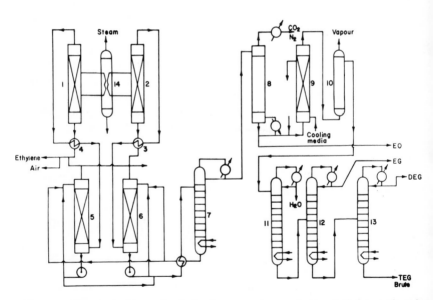

Fig. 28 Nippon Shokubai Kagaku Kogyo Co. process for manufacture of ethylene oxide and glycols.
1 and 2—catalytic reactors; 3 and 4—heat exchangers; 5 and 6—absorbers; 7 and 8—stripper; 9—hydration reactor; 10—evaporator; 11, 12 and 13—dehydration columns; 14—steam boiler for heat recovery and temperature control of reactors. EO—ethylene oxide; EG—ethylene glycol; DEG—diethylene glycol; TEG—triethylene glycol.

glycol at the same installation.[226] A variation in the oxidation cycle is that it is carried out in two steps as depicted in the flow diagram of Figure 28.

Apart from the improved selectivity which is obtained by altering the chemical kinetics of this oxidation process by reducing the concentration of ethylene oxide in the reactor as described earlier, in the Nipon process further improvements are realised by the manner in which the heat of reaction is removed. This is represented schematically in the figure by boiler 14. The cooling circuits for the two reactors are coupled to this boiler in such a manner as to maintain thermal balance within the system. The scrubbing, desorbing, degassing and purification steps of this process are similar to those already described. Thus the only feature of this process which requires further amplification is that of glycol formation. This is relatively simple in as much as the ethylene oxide when hydrolysed produces mono-, di- and triethylene glycol. The glycol solution from the hydration reactor 9 is evaporated in column 10 and fractionated into the respective components by columns 11, 12 and 13.

The Shell Development Co. process produces ethylene oxide and mono-, di- and triethylene glycol using procedures similar to that of the Nipon process.

The Chemische Werke Huels A.G. process on the other hand, utilises the two step oxidation process as described earlier and applied in both the Nipon and Shell processes, but unlike the latter two processes, produces only high purity ethylene oxide. The aldehyde content of the latter has been reported to be less than 50 parts per million. The following material and energy requirements per ton of ethylene oxide have been quoted for this process; 0·96 tons ethylene (at 1·5 atmospheres) 7000 m^3 air (at 6 atmospheres) 150 kWh (at 500 V) 500 kWh (at 6000 V) and 4·9 tons steam (at 20 atmospheres).

3.3 ACRYLIC ACID AND ACRYLIC ESTERS

3.3.1 General Observations

Acrylonitrile is perhaps the most important derivative of acrylic acid. It and methacrylonitrile which is of lesser importance are treated in Chapter 5. In this section the discussion is confined to the chemistry and technology of acrylic acid and its esters.

It is estimated that in 1966-67 about 330-360 000 tons per year of acrylic monomers were produced in the world. The production rates in thousands of tons for various countries were approximately as follows:

USA 184, France 36, West Germany 60, Japan 32, England 21, Canada 10, and Italy 2.

By 1970-71 the worldwide production had reached 400-450 000 tons per year indicating that an annual growth rate of about 5 % was sustained over this period.

The main acrylic monomers which make up this total production are (1) acrylic acid and the acrylates of ethyl, butyl, isobutyl, hexyl, etc. alcohols, and (2) methacrylic acid and the methacrylates of methyl, butyl, isobutyl, hexyl, heptyl, lauryl, stearyl, etc. alcohols. The major uses for these monomers are given in Table 10 along with the approximate percent utilisation for the various applications on a worldwide basis.

TABLE 10
Major uses of acrylic monomers

Application	Approximate utilisation (%)
acrylic fibres	40
finishing agents and binders for fibres	25
dyes	20
sedimentation agents, pastes	5
paper, leather, binders, adhesives, plasticisers, etc.	10

The market for acrylic monomers developed extremely rapidly between 1967 and 1970 due to two factors. The first lies in their special properties which permit their usage to resolve diverse problems of various industries and the second is that the improvements in the manufacturing processes reduced their cost considerably. The latter developments therefore will be examined in some detail.

3.3.2 Processes for Production of Acrylic Acid and Acrylic Esters

The *Reppe process* was applied by Badishe Anilin and Soda Fabrik A.G. for the first time in 1956. This process is based on the synthesis of acrylic acid from acetylene, water and carbon monoxide. Generally the major products from an installation are the methyl and ethyl esters. The overall reactions for a plant process based on the Reppe Synthesis are as follows:

$$CH{\equiv}CH + CO + H_2O \xrightarrow{\text{Ni}} CH_2CHCOOH \xrightarrow{\text{ROH}}$$
$$CH_2CHCOOR + H_2O$$

The synthesis step is carried out at 225°C and 100 atmospheres in tetra-hydrofuran solvent in the presence of certain catalysts. The conversion of reactants per pass is close to 100%, hence no difficult separations are required for the reactor effluent which exits as about an 18% acrylic acid solution in tetrahydrofuran. The acid, as stated earlier, is generally not recovered but is esterified continuously using up to 100% excess of the reacting alcohol. To prevent or reduce polymerisation reactions stabilising agents are added at various points in the system.[235]

This process is utilised also by Rohm and Hass and Dow Badishe in the USA, Lenning in England and Japan Acrylic Chemical Co. in Japan.

The major disadvantages of this process rest with the rather drastic reactor conditions under which the use of acetylene is extremely hazardous. The high cost of acetylene is another factor although this can be reduced somewhat by producing all of the reactants by partial oxidation of methane (i.e. $3CH_4 + 3O_2 \rightarrow CH\equiv CH + CO + 5H_2O$). In the improved Reppe Process the synthesis is effected in one step by utilising an alcohol *in lieu* of water and a nickel carbonyl-hydrogen chloride catalyst system.

In 1966 about 70% of the acrylates produced were obtained by this process. In 1969, the direct oxidation of propylene became competitive and since that time no new investments have been made on the former process. The Reppe process still accounts for about 160-170 000 tons per year of acrylates on the world wide basis.

The Union Carbide Co. (USA) process based on ethylene accounts for about 20 000 tons of acrylates per year. This process involves a substantial loss of nitrogen value (HCN) as well as H_2SO_4 and is unlikely to become more favoured in future. The basic reaction steps of this process are

$$CH_2=CH_2 \xrightarrow{O_2} \underset{\displaystyle CH_2-CH_2}{\overset{\displaystyle O}{\diagup\diagdown}} \xrightarrow{HCN} \underset{\displaystyle CH_2-CH_2}{\overset{\displaystyle OH \quad CN}{|\qquad|}} \xrightarrow[H_2SO_4]{ROH}$$

$$CH_2=CHCOOR + NH_4HSO_4 + H_2O$$

Note that the last step includes simultaneously an hydrolysis, esterification and dehydration step.

The Gooderich Co. process is applied at present by Celanese Chem. Co. and Gooderich Chem. Co. in the USA and Chemcell Ltd in Canada with respective production levels in tons per year of about 16 000; 2500 and 10 000. Either acetone or acetic acid, both of which are low cost chemicals, may be used as the primary raw material in this process. The preference is based primarily on price and is generally slight due to the low price differential between these compounds. A disadvantage of this process is that $AlCl_3$ is required as a catalyst in the oxidation of the intermediary ketene product.

The Ungine-Kuhlmannon process based on the hydrolysis of acrylonitrile[192] became competitive with other processes only after the production of acrylonitrile by the ammoxidation process was substantial enough to reduce the price of acrylonitrile. The hydrolysis and esterification of the nitrile can be carried out simultaneously in the presence of sulphuric acid. Methyl and ethyl esters are generally prepared in this way whereas higher molecular weight esters are prepared by transesterification according to the equation,

$$CH_2{=}CHCOOR + R'OH \overset{H_2SO_4}{\rightleftarrows} CH_2{=}CHCOOR' + ROH$$

About 6000 ton per year of acrylic esters are produced by this process in Japan (Mitsubishi Chemical) at present and about 8000 tons per year in France. The chief disadvantages of this process are the high consumption of H_2SO_4 (about 2 tons per ton of acrylic acid) and considerable wastage of ammonia which is required to neutralise the excess sulphuric acid.

The propylene process was brought on stream by Shell Chemical Corp. for the first time in 1959-60. At the present time it accounts for the production of about 90 000 tons per year of acrylic acid in the USA, and 42 000 tons per year in Japan. It also is used in France and West Germany but production figures are not available to the authors.

The propylene process has many variations. For example Shell converts propylene to acrolein in the vapour phase as a first step, separates the acrolein from unreacted propylene and then converts the latter to acrylic acid in the second step by catalytic oxidation. Sohio, Distillers and Montecatini use the Shell process, also.

A second alternative exists which differs from the first primarily in that the propylene-acrolein separation is omitted and in the second step the mixture of propylene-acrolein is catalytically oxidised to acrylic acid. This route has been adopted by Union Carbide, Mitsui Rayon and Toyo Soda.

In the third alternative propylene is catalytically oxidised directly to acrylic acid in a single phase and single reactor. Gooderich, Japan Catalytic Chemical Industry and Nippon Shokubai utilise this process.

The latter two methods, in particular the last, offer distinct advantages not the least of which are high conversions (100%) and yields (60-70%), long catalyst life and minor amounts of secondary products.

In 1970-71 the production capacity of acrylic acid plants based on the propylene process was about 150 000 tons per year. It is the most important method used on a worldwide basis, today.

3.3.3 Kinetic Considerations

REACTIONS AND MECHANISMS

The main and secondary reactions in the oxidation of propylene are:

1 $CH_2=CH-CH_3 + O_2 \rightarrow$

$$CH_2=CHCH=O + H_2O + 88 \text{ kcal/mol}$$

2 $CH_2=CHCH=O + \frac{1}{2}O_2 \rightarrow CH_2=CHCOOH$

3 $CH_2=CH-CH_3 + O_2 \rightarrow CH_3CH=O + CH_2=O$

4 $CH_2=CH-CH_3 + \frac{1}{2}O_2 \rightarrow CH_3CH_2CH=O$

5 $CH_2=CH-CH_3 + 2O_2 \rightarrow CH_3COOH + HCOOH$

6 $CH_2=CH-CH_3 + 3O_2 \rightarrow 3CO + 3H_2O + 260 \text{ kcal/mole}$

7 $CH_2=CH-CH_3 + 4.5O_2 \rightarrow$

$$3CO_2 + 3H_2O + 460 \text{ kcal/mole}$$

According to researchers who used C_{14} to study the reaction mechanism of propylene oxidation over catalysts comprised of 0.1-1% Cu on SiC, the formation of acrolein (reaction 1) and carbon dioxide (reaction 7) occur simultaneously.[193] Other authors however suggest the sequence propylene $\rightarrow$ acrolein $\rightarrow$ carbon dioxide. The labelled carbon atom of propylene was found in the carbonyl of acrolein.[195,196]

In addition to the secondary reactions listed above cracking reactions also occur which produce carbon oxide (COX) deposits on the catalyst. These deactivate the latter both by reducing the active surface available and by chemically reducing the active oxide surface. This effect is remedied in part by injecting water vapour into the reactor to reduce the amount of carbon formation. An added advantage of water vapour additions to the reactor results from the latter's ability to raise the explosive limit of the propylene-air mixtures. The major technological problem in operating this process arises from the high heat release rates (3500-4000 kcal/kg acrolein) which make it difficult to establish and maintain optimum isothermal conditions in the reactors.

KINETICS

The following observations on the effect of temperature and contact time on the conversion, productivity and yield of acrylic acid from propylene have been reported. They are helpful in indicating the nature of the kinetics of propylene oxidation but are inadequate to describe the system even semi-quantitatively.

(a) In the temperature range 380-480°C the total conversion of propylene increases from 48-50 % to 80-82 % and the conversion of propylene to acrolein increases from 10 % to 50 % when $C_3H_6 : O_2 : H_2O$ mixtures in ratio of 1 : 2·7-2·8 : 2-3 are contacted with a bismuth phosphomolybdate catalyst for 3·8 to 4 seconds. In this same temperature range the variation in conversion of propylene to CO_2 was appreciably less. It increased from about 10 % to about 15 %. However above 480°C the increase in conversion to CO_2 was noted to be much more rapid. At about 475°C the acrolein : formic acid ratio and acrolein : acetaldehyde ratio respectively were noted to be about 8 : 0·8-1 and 7-9 : 1. In general it appears therefore that it is preferable to operate at higher temperatures to achieve higher productivity albeit at some reduction in yield. For example the productivity at 400°C is almost double that at 360°C while the yield loss is only several percent.

(b) The influence of contact time (in the interval 0·8-1·4 sec.) on the conversion, yield and productivity of this system at 390°C and 4-4·5 atmospheres is shown in Figure 29. It can be seen from Figure 29 that the percentage yields of acetaldehyde and formaldehyde increase slightly as the contact time is increased from 0·1 to 1·4 seconds. The total conversion of propylene, total yield of oxidised products and yield and productivity of acrolein, on the other hand, decreased significantly with increasing contact time. A threefold increase in contact from 1·5 to about 4·5 seconds increases the total conversion of propylene (at 480°C) from about 60 % to about 85 % but as stated earlier also increase the conversion to CO_2 thereby reducing the yield of acrolein.

(c) The effect of pressurising the reactor system is twofold. Higher pressure favours increased productivity and increased absorption of acrolein in water. The concentrations by weight of acrolein-water solutions which are obtained from a gaseous mixture containing 2·5 volume per cent acrolein are 2, 8 and 12 % respectively at 1, 4 and 7 atmospheres pressure. At pressures above 4-4·5 atmospheres, condensation of aldehydes is appreciable, and hence operation of the reactor system above this pressure is not necessarily advantageous.

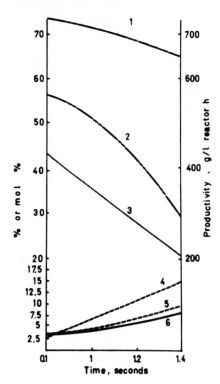

Fig. 29 *Effect of catalyst contact time on the yield, productivity and conversion of propylene by direct oxidation.*

Reactor temperature 390°C, reactor pressure 4-4·5 atmospheres. Graph 1—yield (mol %) of total oxidation products; 2—yield (mol %) of acrolein; 3—productivity (g/l reactor h) of acrolein; 4—yield (%) of acetaldehyde; 5—yield (%) of formaldehyde; 6—total conversion (%) of propylene.

(d) Since the experimental data for the time variation of acrolein yield with respect to concentration of propylene can be fitted to the first order kinetic equation under at least some of experimental conditions studied, it may be tentatively concluded that the production of acrolein is first order with respect to propylene.

CATALYSTS

There are two basic types of catalysts which provide good selectivity for the partial oxidation of propylene to acrolein.

The first type is exemplified by copper oxide on silicon carbide,[241] copper oxide on glass,[242] copper oxide on silicon carbide activated with palladium, rhodium, etc.[243] copper oxide on silica gel with additions of selenium[244] copper oxide with molybdenum, chromium, tungsten and uranium additives[245] and copper oxide on silica gel with additions of beryllium.[246]

The second category of catalysts is based on oxides of metal other than copper. Some examples are oxides of tin and antimony,[247] antimony oxide with tungsten, copper titanium and tin additives,[247] tungsten oxide with additions of silver, tin and bismuth,[248] oxides of antimony, vanadium, bismuth and phosphorus deposited on car-borundum,[249] oxides of tellurium, molybdenum and phosphorus on silica gel,[250] and oxides of molybdenum, bismuth and phosphorus on silica gel.[251]

The latter type of catalyst based on the phosphomolybdate of bismuth is probably one of the best since it also is used in the ammoxidation of propylene.[252]

In the manufacture of acrylic acid without the separation of acrolein (in either single or two phase operation) the following catalysts have been mentioned; molybdates of cobalt,[253-255] of chromium,[256] and of bis-muth,[257] phosphomolybdates,[258] as well as catalysts based on tin, antimony and molybdenum,[259] and on copper and vanadium.[260]

The performance of various catalysts used in the manufacture of acrylic acid by the three methods described earlier are tabulated for easy com-parison in Table 11.

Wakabayashi et al. have studied the activity of the fifth catalyst system listed in Table 11 with a view to correlating the activity to its physical-chemical properties.[261] This study was confined to the conversion of the propylene to acrolein. The following observations and conclusions were drawn.

(a) The electrical conductivity of the binary oxide system of tin and antimony increases significantly as the antimony content increases to about 3 atom per cent. Above this concentration the conductivity decreases slowly. The surface area of this catalyst system also increases up to 3 atom per cent Sb. It was concluded these phenomena can be explained by the formation of solid solutions. The solubility of Sb_2O_5 in SnO_2 depends on the calcination temperature and its duration but is limited by the com-position. For example when an oxide mixture with a $Sn:Sb$ ratio of $3:1$ was calcined at $1000°C$ for 3 hours, X-ray analysis revealed that a large portion of the Sb_2O_5 did not dissolve in the SnO_2. The electrical con-ductivity data indicated that Sb_2O_5 dissolves totally in SnO_2 only up to about 3 atom per cent Sb.

TABLE 11

Performance characteristics of catalysts used in acrylic acid synthesis

Type of catalyst (process)	Temp. (°C)	Contact time (sec.)	Reactant composition	Conversion per pass	Yield (Y) (%) selectivity (S) (%)
1 CuO on Al_2O_3, silica gel or pumice with Se (Distillers) (with acrolein separation)	325-350	2-2·5	2% C_3H_6	90-95	$Y = 81\text{-}82$
2 Biphosphomolybdate on salycilic acid (Sohio) (with acrolein separation)	430-480	1·5-1·6	$C_3H_6:O_2:H_2O$ 1:1·9-2:3·5	56	$Y = 70$
3 Cu covered coils (with acrolein separation)	390-393	0·5	$C_3H_6:O_2$ 1:0·2	78	$Y = 91$
4 Cu_2O on SiC or Al_2O_3 Iodine promoter (Shell) (with acrolein separation)	375-400	0·2	$C_3H_6:O_2:H_2O$ 1:0·2-0·25:1	14	$Y = 68\text{-}81$
5 Sn and Sb oxides—Sn:Sb = 97:3 Sn:Sb = 97:3 Sn:Sb = 3:1 Sn:Sb = 3:1		0·5 2·5 7·5 25			$S = 32$ $S = 17$ $S = 74$ $S = 68$
6 5-10% As oxide 10% Nb As oxide 20% Mb As oxide on silica (Toya Soda)	330-370	(reactor 1)	$C_3H_6:O_2:H_2O$ 1:1·4:12	90	—
7 Mn-Sn-Co-Te (Nippon Shokubai) (single phase)	280-300 400-500	(reactor 2) 7-8	1·9-2% C_3H_6	>20 100	$Y = 50$ $Y = 62\text{-}63$
8 Mo-Mn-Te-P (Gooderich) (single phase)	400-500	54	—	100	$Y = 69\text{-}70$
9 Mo:W:Te:Fe, or (Ni, Cu, Mn) = 1:20:0·01-20:0·001-1:1 (single phase)	320-350	1-20	$C_3H_6:O_2:H_2O$ 1:1·5:1-10		

(b) The rate of acrolein formation at 470° on these catalysts wa
observed to increase sharply to about 5 atom per cent Sb. It appears tha
the rate of formation of acrolein may be dependent upon the amount o
solid solution formed at the catalyst surface. This is supported by the
observation that the activation energy for the formation of CO_2 remain
invariant with catalyst composition up to the atomic ratio of Sn:Sb = 3:
as well as the fact that the selectivity for acrolein (at about 10% conversio
of propylene) increases with additions of Sb_2O_5 but becomes practically
constant at the Sn:Sb ratio of 3:1. The effect of contact time on the
selectivity of two catalyst compositions is shown in Table 11.

(c) The authors proposed that antimony ions are active for propylene
adsorption and that oxygen ions formed at the surface of the catalyst are
active in the formation of acrolein.

The catalyst system listed last in Table 11 can be prepared by heating
mixtures of salts of the metals shown at 200-500°C.[262] It was stated tha
the yields obtained with these catalysts are greater than those realised with
other known catalysts based on Mo-Fe or on W-V-Mo-etc. and that their
effective lives are greater than for the W-V-Mo-Te etc. type catalysts.

3.3.4 Technological Aspects

The separation and purification of acrolein from propylene prior to it
oxidation to acrylic acid can be realised by fractional distillation and/o
by extractive distillation with furfurol or with water.[263,264] The com
position obtainable by distillation using one column is approximatel
(by weight) 80-90% acrolein, 3-10% acetaldehyde, 0·5-3% propion
aldehyde and 2·5% acetone. Propionaldehyde whose boiling point is close
to that of acrolein (49 and 52·5°C respectively) can be separated more
effectively by extractive distillation with furfurol and water.[265,266] The
main impurity of the final purified acrolein is water ($\sim 0·4$%). The level o
other impurities generally does not exceed 0·1%.

It has been stated earlier that a major technological problem of thi
process rests with the requirement to remove the large quantities of hea
released by the oxidation reactions. Although the fluidised bed reactor ha
excellent heat transfer characteristics it suffers in this process, as in others
from the disadvantage of high catalyst losses.[267]

The Toya Soda Manufacturing Co. Ltd process for continuously
producing acrylic acid is shown schematically in Figure 30. This process a
stated earlier is used by Union Carbide and Mitsui Rayon. The basic
features of the process are that the oxidation is carried out in two steps and

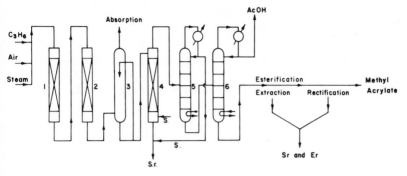

Fig. 30   Toya Soda Manufacturing Co. process for acrylic acid and ester manufacture.
1 and 2—catalytic reactors; 3—condenser and gas-liquid separator vessel; 4—extraction
column (methanol as solvent generally); 5 and 6—distillation columns; ACOH—acetic acid;
S—solvent; E—ester; r—recovered.

two reactors without intermediary separation and purification of
acrolein.[268-271] The reactant mixture comprised of C_3H_6-O_2-H_2O in
the ratio 1:1·4:12 is oxidised in reactor 1 at an average temperature of
about 350°C. The products from this reactor are passed directly to the
second reactor 2 where the acrolein formed in the first reactor is oxidised to
acrylic acid at 280-300°C. The volumetric space velocities in the two
reactors range between 1000-2000 h^{-1} and 2000-4000 h^{-1} respectively.
The products from the second reactor are treated in the absorber 3 to
produce about a 20% acrylic acid solution in water. The acid after
extraction with methanol in the extractor 4 is separated from the
methanol solvent in column 5 and from acetic acid in column 6. The
purified acid from column 6 is then esterified as required by treatment
with the requisite alcohol and concentrated sulphuric acid.

The cost figures for the Toya-Soda process for a production level of
30 000 ton/year of acrylic acid have been quoted to be as follows:[272]

Total cost per ton of acid $114·00 with the main components being
$50·00 for C_3H_6, $16·00 for steam $2·40 for electrical energy, $6·00
for maintenance and $30·00 for depreciation and interest.

The propylene cost quoted above is based on the value of $55·00 per ton
of 90% propylene and the known requirement of 0·88 tons propylene per
ton of acid. If these data are correct the Toya-Soda process produces
acrylic acid 10-15% cheaper than the Reppe process.

CHAPTER 4

Catalytic Dehydrogenation of Alkyl Benzenes and Alkyl Alkanes

4.1 DEHYDROGENATION OF ALKYL BENZENES

4.1.1 General Information

The dehydrogenation processes of most significance commercially will be dealt with in this section. Among these are the manufacture of styrene, divinylbenzene, methylstyrene and m-isopropenyltoluene from ethyl benzene, diethyl benzene, isopropylbenzene, isopropyl toluene and isopropyl-m-xylene.

(a) The styrene from ethyl benzene process is of special importance because styrene is one of the monomers which is most widely used in the manufacture of petrochemicals and of plastics. The world production in 1971 was about 4 000 000 tons per year of which about half was produced in the USA.[273,274]

The major uses for styrene are about 60% for polystyrene, 24% for butadiene-styrene rubber and the remaining 16% in a variety of copolymeric materials and other specialty products.

The liquid phase oxidation process, previously mentioned, in which ethyl benzene and propylene are simultaneously oxidised to styrene and propylene oxide respectively may be expected to grow appreciably. At Puertollono, Spain, an installation based on this process currently produces about 60 000 tons per year of styrene and 24 000 tons per year of propylene oxide.

The main factor which determines the price of styrene is the cost of the raw material which accounts for about 70% of all costs. Benzene as raw material would account for about 50% of the total costs. The operating expenses of the processes vary considerably. For example in old installations steam requirements are about 13 tons per ton of styrene, whereas in new installations this has been reduced to about one half of this amount.

(b) Divinylbenzene from diethylbenzene also is an important process because of the special properties and hence wide usage of ion-exchange resins prepared from copolymers of styrene and divinylbenzene.

(c) α-methyl styrene from isopropyl benzene is the only process used today to produce this chemical on a commercial scale. The α-methyl styrene is used as a substitute for styrene in the manufacture of butadiene-styrene rubbers.

The primary reason for the success of the latter process rests with the fact that isopropyl benzene (cumene) is produced in large quantities (hundreds of thousands of tons per year) for the simultaneous manufacture of phenol and acetone in accord with the following basic reactions:

Benzene + propylene $\rightarrow$ cumene $\rightarrow$ cumene hydroperoxide $\rightarrow$

phenol + acetone

The propylene raw material for the alkylation step may contain up to 70 % propane without serious impairment to this reaction.

Another transformation which the cumene hydroperoxide may undergo is to α-methyl styrene by dehydration of the intermediate benzene isopropanol.

The alkylation processes for benzene and the dehydrogenation (dehydration) reactions to produce styrene and α-methyl styrene have many features in common. Hence the production of phenol, acetone and α-methyl styrene within a single installation should be particularly attractive. The methyl styrene produced by the latter transformation of cumene hydroperoxide may be rehydrogenated to cumene for recycle to the phenol-acetone plant or combined with dehydrogenated products of cumene and fractionated to produce purified α-methyl styrene.

(d) Processes of dehydration of isopropyl toluene and isopropyl-*m*-xylene also present many aspects which are similar to the manufacture of styrene and α-methyl styrene. The isoalkenyl aromatic compounds are widely used in the production of thermoplastic and plastic materials as well as synthetic rubbers.

4.1.2   Thermodynamic, Kinetic and Technological Considerations

DEHYDROGENATION OF ETHYL BENZENE

Reactions and Thermodynamics

The catalytic dehydration of ethyl benzene occurs at 550-660°C in accord with the following equation:

$$C_6H_5CH_2CH_3 \rightleftarrows C_6H_5CH=CH_2 + H_2$$

$$\Delta H = 28 \cdot 7 \text{ kcal/mol}, \qquad \ln K_p = 15 \cdot 6 - 14\,900/T°K$$

Secondary reactions which do not reach equilibrium include the following

$$C_6H_5CH{=}CH_2 \rightarrow C_6H_5{-}C \equiv CH \text{ (phenyl acetylene)} + H_2$$

$$2C_6H_5CH{=}CH_2 \rightarrow C_6H_5CH_2CH_3 + C_6H_5C{\equiv}CH$$

$$C_6H_5CH_2CH_3 \rightarrow C_6H_6 + C_2H_4 \qquad \Delta H = +24{\cdot}3 \text{ kcal/mol}$$

$$C_6H_5CH_2CH_3 + H_2 \rightarrow C_6H_6 + C_2H_6 \qquad \Delta H = -10{\cdot}0 \text{ kcal/mol}$$

$$C_6H_5CH_2CH_3 + H_2 \rightarrow C_6H_5CH_3 + CH_4 \qquad \Delta H = -15{\cdot}4 \text{ kcal/mol}$$

$$8C + 16H_2O \rightarrow 8CO_2 + 16H_2 \qquad \Delta H = +190{\cdot}0 \text{ kcal/mol}$$

Because of the secondary reactions the effective heat of reaction is 33·2 kcal/mol of styrene produced.[275] The equilibrium constant for the main reactions may be evaluated from the thermodynamic data given in Tables 12 and 13. Several values of the equilibrium constant[276,277] calculated for different temperatures are listed in Table 14.

TABLE 12
Thermodynamic data for ethyl benzene and styrene at 298·16°K

Component	$\Delta H_c^\circ (kcal/mol)$	$\Delta H_f^\circ (kcal/mol)$	$\Delta F_g^\circ (kcal/mol)$	ΔS_g° calories per degree
Ethyl benzene (gas)	1101·13 ± 0·17	7·12 ± 0·20	31·21	−80·79
Styrene (gas)	1060·90 ± 0·20	35·22 ± 0·24	51·09	−53·24

TABLE 13
Values of ΔH and ΔF for ethyl benzene and styrene at various temperatures

Temperature (°C)	Ethyl benzene (kcal/mol)		Styrene (kcal/mol)	
	ΔH_f°	ΔF_f°	ΔH_f°	ΔF_f°
327	2·488	57·696	31·81	68·67
427	1·529	66·921	31·08	74·87
527	0·798	76·302	30·51	81·16
627	0·266	85·779	30·09	87·53
727	−0·061	95·303	29·83	93·92

TABLE 14
Equilibrium constants for ethyl benzene-styrene reaction

Temperature (°C)	H° (kcal/mol)	F° (kcal/mol)	log K_p	K_p (atmospheres)
327	29·322	11·024	−4·0160	$9·65 \times 10^{-5}$
427	29·551	7·949	−2·4817	$3·30 \times 10^{-3}$
527	29·712	4·858	−1·3269	$4·71 \times 10^{-2}$
627	29·824	1·751	−0·4252	$3·75 \times 10^{-1}$
727	29·891	−1·383	−0·3022	$3·00 \times 10^{0}$

Chemical Kinetics

The kinetics of the main reaction have been described by the following form of equation :[278,277]

$$\frac{d(ST)}{dw} = k\left(P_{EB} - \frac{P_{ST}P_{H2}}{K}\right)$$

where

　　d(ST) is the styrene made (kmol/h) in element of volume continuing
　　　　dW kg of catalyst
　　K is equilibrium constant
　　k is reaction rate constant kmol/h atmos kg catalyst
　　P_{H2}, P_{ST}, P_{EB} respective partial pressures (in atmospheres)

According to Balandin et al.[280] there is a structural correlation between the reactant molecules and the crystalline lattice of the catalyst. The reaction rate constant (k) may be calculated from experimental data using the following equation model :

$$k_1 = 2·303(Z_1 + Z_2)A_1 \log\frac{A_1}{A_1 - m} - m(Z_1 + Z_2 - 1)$$

where

　　A_1 is amount of reactant passed over catalyst per unit time and m is
　　　　amount of product formed
　　Z_1 and Z_2 are relative adsorption coefficients of the reaction products,
　　　　styrene and hydrogen, as calculated from the equation

$$Z = \frac{(m_0/m) - 1}{(100/P) - 1}$$

where m_0 and m are quantities of the product formed when the feed consists of pure ethyl benzene and when it contains P percent of the reactant in the reaction products for which Z is determined.

On the basis of this equation model and acquired experimental data they calculated the reaction rate constant to be 0.370×10^{-2} gmol/min at 530°C and 0.780×10^{-2} gmol/min. at 545°C.

Technological Kinetic Considerations

The effects of initial temperature of the ethyl benzene, water content and reactant mixture composition, as well as of the back mixing of reaction products, on the conversion values are of special technological importance.

The largest part of styrene production is obtained by dehydrogenation in adiabatic reactors at 580-630°C at atmospheric pressure using catalysts with an iron oxide base and chromium, potassium and various other additives.[281] To obtain increased yields of styrene, superheated water vapour at 710-730°C (in amounts of 2.6-3.6 kg/kg of ethyl benzene) is added to the reaction zone. The water vapour injection also aids in the regeneration of the catalyst. The conversion of ethyl benzene is about 45-60 %; the selectivity of the process is 84-90 % for a space velocity of the ethyl benzene 0.25 to 0.7 h^{-1}.

To make the adiabatic process competitive with the isothermal process it was necessary to increase the conversion, selectivity and productivity per unit weight of catalyst and to reduce the consumption of water vapour.

The following observations have been made using a continuous adiabatic reactor with the following operating limits: initial temperature of ethyl benzene 480-600°C, water vapour temperature 710-870°C, mixing zone temperature 600-660°C and contact time 0.1-0.2 seconds.

At temperatures below 600°C the decomposition of ethyl benzene to benzene is negligible as shown in Figure 31. The degree of decomposition to both benzene and styrene increases with temperature, the former less slowly than the latter. However there also is strong dependence upon the ratio of water vapour to ethyl benzene for these decompositions.

The effect of preheating the ethyl benzene up to 600°C on the conversion is shown in Figure 32. It can be seen that this effect is relatively more significant on conversion to styrene than to benzene.

Temperature increases in the mixing zone lead to rapid increases of ethyl benzene decomposition with considerable increases in conversion to benzene as shown in Figure 33.

Mixtures of ethyl benzene and water vapour with initial temperatures of 540 and 840°C respectively decompose less than those with initial temperatures of 600 and 730°C respectively. This is in agreement with the caloric content of the two mixtures which as determined from the graphs of Figure 34 are 837 and 801 kcal/kg respectively. Figure 34 may be used, also, to calculate the temperatures of the superheated vapours introduced into the mixing zone and of the ethyl benzene-water vapour mixture at the exit to this zone. In this way the heat losses in the catalytic zone can be evaluated.

The determining factor in the decomposition of ethyl benzene is the initial temperature of the mixture. Preheating the ethyl benzene to 600°C ensures a conversion to benzene of less than 1 %. This is indicated in Figure 32 and shown by the graphs in Figure 35.

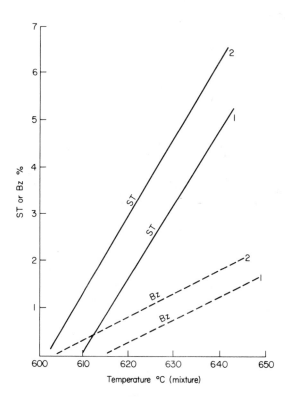

Fig. 31   Variation of styrene and benzene content as a function of the decomposition temperature
of reactant mixture and the ratio of water vapour/ethyl benzene.
1—H₂O/EB = 1 ; 2—H₂O/EB = 3.

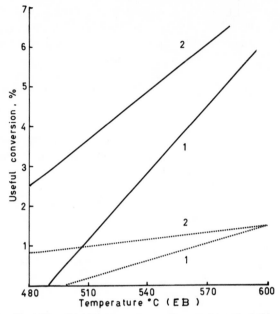

Fig. 32   Effect of initial temperature on decomposition of ethyl benzene.
—— Conversion to styrene; ···· Conversion to benzene at $H_2O/EB = 2$ and $H_2O(v)$; Temperature for graph 1 = 730°C and graph 2 = 840°C.

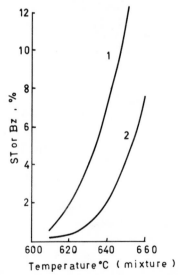

Fig. 33   Effect of mixing zone temperature on decomposition of ethyl benzene.
1—decomposition to styrene; 2—decomposition to benzene.

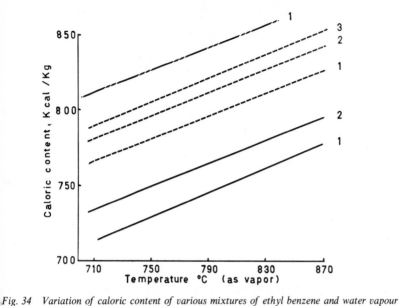

Fig. 34   Variation of caloric content of various mixtures of ethyl benzene and water vapour as a function of the initial temperature of the water vapour.
Initial temperatures for ethyl benzene are 480, 560 and 600°C for graphs 1, 2 and 3 respectively and water to ethyl benzene ratios are as follows:
2 for graphs 1 and 2 ———, 3 for graphs 1, 2 and 3 – – – – and 4 for graph 1 – – – – – – –.

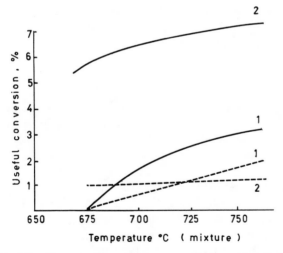

Fig. 35   Effect of reactant mixture temperature on decomposition of ethyl benzene.
——— conversion to styrene; – – – – conversion to benzene; 1—E.B. temperature—480°C; 2—E.B. temperature = 600°C.

The conversion values and caloric content of mixtures of ethyl benzen and water vapour can be calculated for the equation

$$C_f - C_i = \frac{(I_i - I_f)(n + 1)}{Q_p}$$

where

C_i and C_f are initial and final conversions %

I_i and I_f are initial and final caloric contents of the mixtures of ethy benzene and water vapour (kcal/kg)

n is the ratio of water vapour to ethyl benzene by weight

Q_p is the caloric requirement for dehydrogenation of ethyl benzen (33 200 kcal/mol)

$$I_{EB} + nI_{H_2O}(v) = (n + 1)I_i$$

Where I_{EB} and $I_{H_2O}(v)$ are caloric contents of ethyl benzene and super saturated water vapour at the entrance of the reactor (kcal/kg).

The value of I_f is determined at the minimum temperature possible fo dehydrogenation of ethyl benzene which is 565-585°V (some authors specif 575°C).

From the foregoing observations it can be deduced that the conversio of ethyl benzene and yield of styrene can be increased and the consumptio of steam and formation of secondary products decreased by reacting th superheated vapours with a water/ethyl benzene ratio of 3:1 at 735 760°C. Direct contact of the ethyl benzene with the superheated wate vapour in zones of maximum temperature should be avoided as much a possible to ensure uniform distribution of reactants throughout th reactor.

Technological Kinetics of Oxidative Dehydrogenation

The dehydrogenation of ethyl benzene by oxidative dehydrogenation i the presence of hydrogen acceptors has been studied extensively. Suc processes with active catalysts permit better yields and selectivities a lower temperatures than those obtained by processes without accept ors.[282-287]

The use of SO_2 (*in lieu* of O_2) as a hydrogen acceptor offers the advantag that its reaction with hydrogen is slightly endothermic and hence can b more easily controlled.[288,289] Yields obtained using SO_2 and a catalys of $Ca_8Ni(PO_4)_6$ in excess of 80 mol% have been reported.[288] Simila yields with high selectivity also have been reported for catalyst of th same type promoted with Cr_2O_3.[289] When SO_2 is used as a hydroger cceptor the reaction proceeds as follows:

$$3C_6H_5CH_2CH_3 + SO_2 \rightleftarrows 3C_6H_5CH=CH_2 + H_2S + 2H_2O$$

he following observations and conclusions have been made from studies f the catalytic oxidative dehydrogenation of ethyl benzene in a continuous ow reactor of the fixed bed type. The catalyst used in this study consisted f $Ca_8Ni(PO_4)_6$ with 1.5% Cr_2O_3 promoter. It had a microporous tructure with most of the pores being about 1000 Å in diameter.

1   A reduction of the partial pressure of ethyl benzene from 1 atmosphere to 0.05 atmospheres permitted the same conversion (about 90 mol %) to be attained at temperature about 160°C lower (900°K vs 1060°K).
2   The introduction of SO_2 as a hydrogen acceptor permitted high conversions to be attained at even lower temperatures (90 mol % at 600°K, see Figure 36).

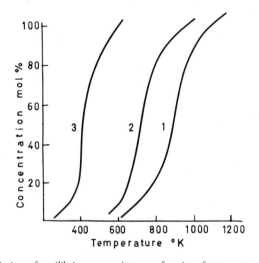

ig. 36   Variations of equilibrium conversions as a function of temperature, partial pressure of ethyl benzene and hydrogen acceptor presence.
raph 1—partial pressure of ethyl benzene = 1 atmosphere; Graph 2—partial pressure of thyl benzene = 0.05 atmosphere; Graph 3—partial pressure of ethyl benzene = 0.75 atmosphere with addition of SO_2.

3   In the temperature interval 500-580°C at molar ratios of $EB:SO_2:N_2$ of $1:1:2$ and space velocities of $0.5-5\,h^{-1}$, conversions of 80-88 % with a selectivity of about 80 mol % were obtained. Further particulars are given in Table 15. The activation energy calculated in

TABLE 15
Variation of conversion, yield and selectivity with temperature and contact time

Temper-ature (°C)	Space velocity (h^{-1})	Contact time (sec.)	Yield (mol %)	Conversion (mol %)	Selectivity (mol %)
520	1·0	1·71	45·0	52·4	86·0
	2·5	0·68	59·2	42·1	92·2
540	0·5	3·3	69·5	88·5	78·0
	4·4	0·38	37·5	40·4	93·1
560	2·0	0·82	53·7	70·1	76·6
	4·2	0·38	44·1	51·0	90·5
580	1·0	0·88	69·0	85·2	61·0
	4·6	0·34	47·9	53·0	89·4

accordance with the Arrhenius equation from these data was foun
to be 18·1 kcal/mol in the presence of SO_2 and 29·25 kcal/mol i
the absence of any acceptor.

4   In the presence of SO_2 acceptor significantly greater conversion
were achieved for the same contact time. For example, at 450°C an
0·6 seconds contact time the conversion in the presence of SO
was 40-41 % compared with 13 % in the absence of an accepto
(see Figure 37).

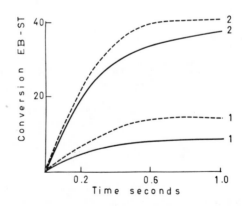

Fig. 37   Conversion of ethyl benzene to styrene as function of contact time.
—— at 500°C; ---- at 540°C; for $P_{EB} = 0.25$ atmospheres; Graphs 1—for EB + N_2
Graphs 2—for EB, N_2 + SO_2.

Catalysts

As in all catalytic processes the dehydrogenation catalysts constitute one of the most important parameters of the dehydrogenation process. Because of the large number of secondary reactions which may occur during dehydrogenation processes the catalyst must have a particular high selectivity and activity and as always good mechanical, thermal and chemical stability.

The selectivity as well as the activity of a catalyst is determined to a large degree by its structure.

The phenomena of sorption, diffusion, specific activity and heat transfer also are determined by the structure of the catalyst in particular its capillary structure. Thus the preparation, indeed the definition of catalysts with optimal structure requires detailed microscopic studies including the use of electron microscopes.[290]

The capillaries of catalysts are classed as macro- and micro-capillaries for convenience, but the distinction is arbitrary and not well defined. Capillaries with radii smaller than 150 Å may be classed as micro-capillaries. The following conclusions have been reached about a series of complex catalyst systems based on compounds of Fe, K, Cr, Ca, Al, Mg, etc.

About 65-75 % of the pore volume is due to pores with radii greater than 150 Å, about 10-20 % are due to pores with radii between 4 and 150 Å and the remainder due to pores with radii less than 4 Å. Thus as an example the catalyst PO-5 (based on Fe-Ca) has been found to have 67·6 % of the pore volume attributable to pores with radii between 150 Å and 7·5 μ, 16 % to pores with radii between 4 Å and 150 Å and 16·4 % to pores with radii under 4 Å. The SK-60 (based on FeCr) pore volume was distributed 72·6 % among pores with radii between 150 Å and 25 μ, 18·4 % among pores with radii between 4 Å and 150 Å and 9 % among pores with radii under 4 Å.

Integral curves of the distribution of macro-capillaries in catalysts as determined by a mercury porosimeter are similar in form, independent of the origin and composition of the catalysts. The SK-60 system after 7000 hours of use showed a slight displacement of the curve toward smaller radii. This demonstrates high thermo-mechanical stability of this catalyst.

Another important problem encountered with dehydrogenation catalysts is that of carbonaceous deposits. These deposits have the general formula $(CH_x)_n$ in which the hydrogen content decreases gradually as a function of the time the catalyst has been used.[291] The amount of deposits formed is directly related to the reaction conditions and can be minimised by proper control.

The intensive abuse to which catalysts in these processes are subjected necessitates proper definition of the parameters which affect the formation of surface deposits. Studies of this type have been described in the literature for several catalysts.[293,294] The following conclusions have been reached in this regard for the catalyst system Fe_2O_3-Cr_2O_3-K_2CO_3 for the dehydrogenation of ethyl benzene in an industrial reactor at a space velocity of $0.5\ h^{-1}$ and a weight ratio of ethyl benzene to water of $1:2.6$. About 80% of the pore volume for the catalyst studied was due to pores with radii greater than 1000 Å. The energy of activation for dehydrogenation of ethyl benzene was found to be 31 kilocalories per mole.

The deposits can be removed by reaction with water vapour via the water gas reaction thereby regenerating the catalyst. The amount of deposits formed can be reduced also by diluting the reactants (ethyl benzene) with inert gas (N_2). Dilution favours the dehydrogenation by reducing the partial pressure of ethyl benzene (see Figure 36). No means was found to completely avoid the formation of deposits or to completely remove them. Consequently, to maintain the activity at a constant practical level a gradual increase in the reaction temperature is necessary.

As an example of the degree of catalyst deactivation produced by these deposits the following particulars are cited. In order to maintain the same conversion (about 40-45% by weight) and the same selectivity (about 88% by weight) after 20, 80 and 125 hours of use it was necessary to operate the

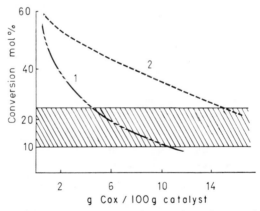

Fig. 38   Variation of conversion as a function of Cox deposit and temperature.
For PEB = 0.5 atmospheres and space velocity = $0.5\ h^{-1}$; graph 1 at 580°C and graph 2 at 620°C. Shaded area represents the thermal conversion zone.

eactor at 590, 600 and 630°C respectively. These increases in the thermal
evel of the reaction occurred under cyclic operation following two regen-
rations. At 580 and 620°C respectively almost total deactivation occurred
it 5 and 15 grams of deposit per 100 grams of catalyst. This effect is
raphed in Figure 38 where the shaded area represents the zone of thermal
onversion.[281]

The deactivation phenomenon can be characterised by an equation of
he form[295]

$$\frac{dT}{dq} = \frac{k_{p0}}{E_a} R T^2 \exp\left(-E_d/RT\right)$$

vhere

$\dfrac{dT}{dq}$ = temperature variation required to effect a constant conversion rate
per unit weight increase of deposit on catalyst surface

k_{p0} = exponential factor (h^{-1})

E_a = activation energy (31 000 cal/mol)

E_d = deactivation energy (cal/mol).

n the foregoing system the deactivation energy was found to increase
ontinuously until the process gradually became a non-catalytic thermal
rocess. After four consecutive cycles the following values for E_d were
alculated.

Cycle	Hours of operation	E_d (cal/mol)
1	1-115	26 306
2	132-227	48 850
3	231-335	72 356
4	1484-1624	103 775

The deactivation is reversible and the initial catalytic activity can be
pproached by regeneration. The catalyst activity, reflected by the con-
ersion value of ethyl benzene (in mol%) is related to the amount of
leposits (q) and may be expressed by an exponential equation of the
Froment-Bischoff type[296]

$$A_c = \beta E^{-\alpha q}$$

vhere

A_c = activity

β, α = constants at particular temperature.

The above equation has been verified by following the variation of catalyst activity[294] as a function of time and temperature of the reaction. The pertinent data are summarised in Table 16.

TABLE 16

Temperature (°C)	Time (h)	$q(g/100\ g$ catalyst)	% Ac. experimental	α	β
580	1·0	1·17	1·653	0·162	54
	7·8	5·75	1·322		
620	0·5	0·8	1·763	0·0528	60
	8·0	14·86	1·477		
etc.					

The time dependence of the amount of deposit formed (grams cox/100 grams of catalyst) can be expressed by the relation $q = ht^n$ where h and n are constants depending upon the nature of the catalyst only.[287] The amount of deposit formed also depends upon the temperature, increasing quite sharply with increase in temperature. For example at 580°C, $q = 1.5$ and 5·5 after 2 and 7 hours operation respectively, whereas at 620°C after 7 hours operation $q = 12.0$.

The ratio of carbon in the deposit $(CH_x)_n$ increases with time of operation and does so more rapidly at higher temperatures. At 620°C the carbon content reaches the maximum value of 95% after about 4 hours of operation.

The catalysts used in the dehydrogen of butylenes are equally effective for dehydrogenation of ethyl benzene. Those based on oxides of Zn, Mg etc. are gradually being replaced by catalysts based on Fe. One of the reasons for this is because Al_2O_3 which is used as a promoter for the Zn type catalysts imposes narrow concentration limits; in small quantities it is effective at high temperatures only whereas in larger amounts it favours secondary reactions.

The conversion per pass of ethyl benzene to styrene with iron based catalysts is significantly higher than with zinc based catalysts. The styrene content of the product using the former type of catalyst may be as high as 50-60% by weight, but only 36-40% with the latter. Thus the productivity of an installation can be increased substantially with the consequent reduction in the utility costs by switching from a zinc based catalyst to one based on iron.

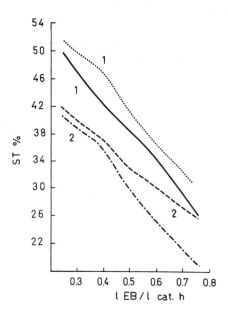

Fig. 39   Effect of ethyl benzene feed rate on the styrene content of its dehydrogenation products.
Graph 1—for catalyst type PO-5; Graph 2—for catalyst type SK. —— at 570°C; ···· at 600°C.

In Figure 39 the performance of two catalysts are compared. It can be seen that the PO-5 type (Fe-Ca) has a better performance than the SK type (Fe-Cr). The mechanical resistances (kg/cm²) pore volumes (cm³/g) and volumetric weights (kg/l) for these catalysts were respectively 285, 0·2 and 1214 and 130, 0·2 and 1168.

The PO-5 type catalyst when used at 624-632° with a feed rate of 1 litre ethyl benzene per litre of catalyst per hour produced a product stream containing about 60 % styrene with a yield of about 94 %.[298]

Iron base catalysts when utilised with high ratios of steam to ethyl benzene require little regeneration. A catalyst comprising 90 % Fe_2O_3, 4 % Cr_2O_3 and 6 % K_2CO_3 with a productivity of 170 g styrene per litre of catalyst per hour has been found to be particularly selective in this regard. The Fe_2O_3-7·7 % ZnO catalyst system on the other hand is less effective and at 566-590°C has a useful operating life of only 1-3 months.

In addition to the above catalysts the following have been reported recently to be effective for this process also; autoregenerative catalysts in electrical heated reactors[299] catalysts based on FeO, Cr_2O_3[300] and special catalysts for oxidative dehydrogenation of ethyl benzene based on Nb[301] or Nb-Cr[302] in the presence of oxygen and based on $CaNi(PO_4)$ in the presence of SO_2.[303]

DEHYDROGENATION OF ISOPROPYL BENZENE

The thermodynamic and kinetic aspects of the dehydrogenation of isopropyl benzene (cumene) to α-methyl styrene are similar to that of the process for the dehydrogenation of ethyl benzene to styrene. The following brief comments are made for completeness and added clarification.

(a) The main endothermic reaction is:

$$C_6H_5-CH(CH_3)_2 \rightleftharpoons \overset{\overset{\displaystyle CH_3}{\displaystyle |}}{C_6H_5C}{=}CH_2 + H_2$$

Besides the process of pryolysis[304] the following secondary reactions also occur:

$$C_6H_5CH(CH_3)_2 \longrightarrow C_6H_6 + CH_2{=}CHCH_3$$

$$C_6H_5CH(CH_3)_2 \longrightarrow C_6H_5CH_3 + CH_2{=}CH_2$$

$$C_6H_5CH(CH_3)_2 \xrightarrow{+H_2} C_6H_6 + CH_3CH_2CH_3$$

$$C_6H_5CH(CH_3)_2 \xrightarrow{+H_2} C_6H_5CH_3 + CH_3CH_3$$

$$C_6H_5CH(CH_3)_2 \xrightarrow{-H_2} \overset{\overset{\displaystyle CH_3}{\displaystyle |}}{C_6H_5C}{=}CH_2 \xrightarrow{+2H_2} C_6H_5CH_2CH_3 + CH_4$$

(b) The degree of conversion is increased (as for ethyl benzene) by reducing the partial pressure of the reactants, by dilution with superheated steam or inert gas (N_2). The degrees of conversion obtained for various molar ratios of steam to cumene are listed in Table 17 for several temperatures. In the case of adiabatic reactors using autoregenerative catalysts or catalyst regeneration every 24 hours with a cumene feed rate of 0·5 litres per litre of catalyst per hour, the performance with a molar ratio of steam to cumene of 3 is satisfactory. The conversion values obtained industrially are about 50% and yields range between 94 and 96%. The equilibrium constant for the main reaction is 0·46 at 520°C and 3·2 at 620°C.

(c) Kinetic studies have been made for both the dehydrogenation of ethyl benzene and cumene on a chromium oxide catalyst.[305] The rate of cumene dehydrogenation is about double that for ethyl benzene. For example at 550°C the respective rate constants found were 0·78 and 1·57

TABLE 17
Variation of conversion of cumene with molar ratio (n) of water
vapour/cumene and temperature

Temperature (°C)	Conversion (mol %)		
	$n = 0$	$n = 10$	$n = 20$
520	0·56	0·87	0·92
540	0·64	0·90	0·94
560	0·72	0·98	0·96
580	0·78	0·96	0·97
600	0·93	0·97	0·98
620	0·99	0·98	0·99

respectively. The activation energy for the dehydrogenation of cumene was determined to be 42·07 kcal/mol.

(d) In addition to the catalysts mentioned earlier those based on oxides of cerium,[306] vanadium[307] and other metals with stable oxides[308] have been proposed. Catalysts based on iron oxide are mentioned with increasing frequency in the recent literature.

4.1.3 Technological Aspects

THE CASE OF ETHYL BENZENE DEHYDROGENATION

A schematic flow diagram for the manufacture of styrene by dehydrogenation of ethyl benzene in an adiabatic reactor is shown in Figure 40.

In this process the ethyl benzene is preheated to 560-570°C by heat exchange with the reaction products in exchangers 3 and 4. The steam which is superheated in the furnace is used to raise the reactor temperature as well as to regulate the water to ethyl benzene ratio. The main control of the latter ratio is effected however by steam at about 6 atmospheres which is premixed with the ethyl benzene prior to the heat exchangers. This low pressure steam is generated at exchanger 5 thus ensuring maximum economy of heat for the process.

The organic phase is separated in unit 6 and distilled in columns 7, 8, 9 and 10. A mixture of benzene and toluene (formed by secondary reactions as shown earlier) is obtained from column 7, unreacted ethyl benzene from column 8 and styrene from columns 9 and 10. The unreacted ethyl benzene is recycled. The conversion is about 40 % per pass and the overall yield is about 90 %.

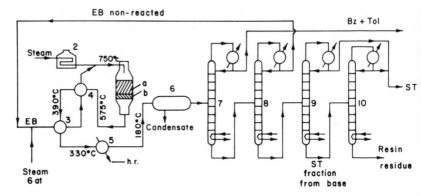

Fig. 40   *General flow diagram for manufacture of styrene by dehydrogenation of ethyl benzene. 1—catalytic reactor, a—catalyst bed, b—filling ring; 2—furnace; 3,4 and 5—heat exchangers; 6—separator; 7,8,9 and 10—distillation columns; h.r.—heat recuperator; E.B.—ethyl benzene; Bz—benzene; Tol—toluene; ST—styrene.*

The ethyl benzene used as feed material must be of high purity in particular with respect to diethylbenzene which at concentration as low as 0·05 % produces sufficient divinylbenzene polymers to cause difficulties in operation. The presence of benzene on the other hand improves the process somewhat and some authors recommend its use as a feed diluent.

The isothermal dehydrogenation of ethyl benzene is carried out in tubular reactors which are heated externally by combustion gases. The consumption of steam in this case is less than for the adiabatic operation, being about 1·2-1·3 versus 2·2-2·3 tons per ton of ethyl benzene. The fuel is supplied in the gaseous phase as recovered from the final condensation of the reaction products. The temperature of the combustion gases at the entrance to the multi-tubular reactor is 695-705°C and falls to 615-620°C at the exit.

The fractionation and purification of products is not free of complications, in particular since the products benzene, toluene and styrene must all be of high purity (99·5-99·7 %). To reduce the degree of contamination by polymeric materials the distillations are carried out under vacuum and inhibitors such as hydroquinone, sulphur or *p*-tert-butyl pyrocatechol are added in the second step distillations. The effect of temperature on the polymerisation rate of styrene is high. At 60°C about 0·1 % polymerisation occurs whereas at 90°C 1·8 % of the styrene polymerises per hour. In separating styrene from ethyl benzene a minimum of 75 to 80 trays are required in the distillation column to effect adequate rectification because the boiling point of ethyl benzene (136°C) is close to that of styrene (145.2°C).

The benzene-toluene column utilises 30-32 trays and operates with a base temperature of 96-97°C and a top temperature of 56-58°C at 175-178 mm Hg pressure. This compares with 75-80 trays with a base temperature of 90°C and a top temperature of 50-52°C at 25-35 mm Hg pressure for the ethyl benzene-styrene column(s).

The styrene obtained upon separation from ethyl benzene contains excessive amounts of inhibitor which is removed by rectification in a packed column with the equivalent of about 1·1 theoretical plates under about 15 mm Hg pressure at the top and a base temperature of 57°C. This final product is used for polymerisations and contains a maximum of 10 ppm p-tert-butyl pyrocatechol, 0·1 % polymers and 0·2-0·3 % ethyl benzene and isopropyl benzene.

The material and energy requirements per ton of styrene are : 1·15 tons ethyl benzene, 1·81 kg of catalyst, about 2 tons of steam at 17 atmospheres plus about 14 tons at 6 atmospheres and 135 kWh of electrical energy.

The processes of significant application on the international level are the Badger, Lummus-Monsanto and Dow.

The Badger Co. Inc. process is characterised by the primary alkylation of benzene with ethylene using an $AlCl_3$ catalyst followed by dehydrogenation of the ethyl benzene. The reaction products are separated as described earlier except that a special single column of high efficiency is used to effect the separation of ethyl benzene and styrene with a reported savings in investment and operating costs. The process is used by Badger, Cosden and Union Carbide. On the world basis this process accounts for about 2 million tons of styrene per year but only about 200 000 tons per year in the USA.[309]

The Lummus Co. and Monsanto Co. process differ from the Badger process in that the reaction products are split into a fraction comprised of unreacted ethyl benzene and toluene and a fraction from which styrene is obtained by further fractional distillations. The process is reported to have a low steam consumption. Among the companies using this process are Monsanto (about 300 000 tons per year) and Montecatini (400 000 tons per year).

The Dow Chemical Co. process[273] uses a steam to ethyl benzene ratio of 2·2 to 1 with a short entrance or retention zone prior to the catalyst bed and alternate directional feed through the latter. The Shell type 105 and 205 (Fe_2O_3-Cr_2O_3-K_2CO_3) catalysts are used except in Germany where the zinc phosphate type is still used. The product separations are carried out as described earlier using sulphur or hydroquinone as inhibitor. About 400 000 tons per year of styrene are produced by this process in the USA.

The dehydrogenation of ethyl benzene is mentioned frequently in the patent literature and recent patents hold some relevance to the processes described.[310-312]

THE CASE OF ISOPROPYL BENZENE DEHYDROGENATION

The technological aspects of α-methyl styrene synthesis are closely similar to those of styrene manufacture. Superheated steam is employed in this process, also to provide the necessary endothermic heat of reaction and control of the conversion reactions. The reaction products are treated similarly except that in this case, ethyl benzene, styrene, benzene and toluene are separated as a first fraction from the isopropyl benzene and α-methyl styrene. The ethyl benzene and styrene are then separated from the benzene and toluene and finally the unreacted isopropyl benzene is distilled off from the product and recycled. The latter is obtained, thus, in a single column with a purity of 99·7-99·8 %. About 1·15 tons of isopropyl benzene are required per ton of α-methyl styrene. The overall process is outlined in Figure 41.

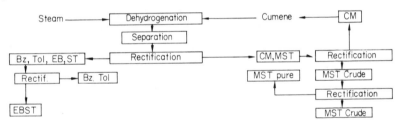

Fig. 41   Manufacture of α-methyl styrene from cumene.
CM—cumene; Bz—benzene; Tol—toluene; ST—styrene; MST—α-methyl styrene; EB—ethyl benzene.

DEHYDROGENATION OF ISOPROPYL TOLUENE, ISOPROPYL m-XYLENE AND DIETHYL BENZENE

The alkenyl aromatics obtained by dehydrogenation of the above two isopropyl derivatives have a wide use in the manufacture of thermoplastic materials and synthetic rubbers. Divinylbenzene resulting from the dehydrogenation of diethylbenzene is used primarily in the manufacture of ion exchange resins.

Figure 42 depicts an installation schematically, which was employed to produce m-isopropenyl toluene (IPenT) and isopropenyl-m-xylene

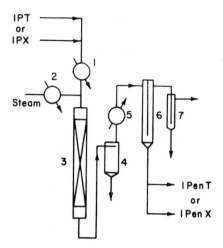

Fig. 42   General flow diagram for a pilot plant process for fixed or fluidised catalytic bed
dehydrogenation of isopropyl toluene and isopropyl xylene.
1—preheater at 300°C; 2—preheater at 580-620°C; 3—reactor; 4—cyclone; 5—condenser;
6—gas-liquid separator; 7—absorber; IPT—m-isopropyl toluene; IPX—isopropyl-m-xylene;
IPenT—m-isopropenyl toluene; IPenX—isopropenyl-m-xylene.

(IPenX) by dehydrogenation of *m*-isopropyl toluene (IPT) and isopropyl-
m-xylene (IPX) respectively using both a fixed and a fluidised catalytic bed
reactor.

The studies made with the above installation were conducted as follows.
The alkyl aromatics (IPT or IPX) were preheated (by heater 1) to 300°C
and fed continuously into the reactor 3 along with water vapour which was
superheated to 580-620°C. The reaction products were passed through the
cyclone 4 to remove entrained solids, cooled in condenser 5 and separated
into gas and liquid fractions in the separator 6. The reactor unlike the
heaters 1 and 2 was heated by combustion gases. A fixed amount (80 kg)
of 'styrene Kontakt' catalyst was used in each of the runs. The volume of
catalyst in the fluidised bed reactor was about 0·1 m³. The bed height was
0·79 m and the reaction zone height extended to 2·3 m. At a linear vapour
velocity of 0·25 m/sec the time of contact was 3·16 seconds.

Under reaction conditions which were considered optimum for the
fluidised bed operation the conversions obtained were about 52% for
IPT and about 44% for IPX for a single pass and 84·4% and 75·6%
respectively for two passes.

The studies made with the reactor in the fixed bed mode encompassed
the following range of parameters, alkane feed rate 0·7–1·5 litres/h, space

velocity 0.25–$4 \, h^{-1}$, molar ratio $H_2O(v)$ to alkane of 2-3 to 1, molar ratio of oxygen (as air) to alkane of 0.1-0.4 to 1 and a temperature range of 560-620°C. The composition of the alkanes used was, for IPT 85% meta, 10% para and 5% ortho and for IPX 87%, 1, 3, 5 isomer and 13% 1, 3, 4-isomer.

The ratio of isomers after dehydrogenation was found to be the same.[313] It also was established that when the dehydrogenation was carried out in the presence of air an increase in conversion of about 7-8% was obtained.

The dehydrogenation technology of diethylbenzene (DEB) is closely similar to that of ethyl benzene dehydrogenation. However because the product divinylbenzene tends to polymerise more readily it is somewhat more difficult to purify than styrene. The purification is carried out similarly, by distillation, in two steps.[314] In the first step all products with boiling points lower than ethyl styrene are separated from the top of the column. This fraction comprises 50-60% of the total feed to the column and contains 40-60% unreacted DEB and ethylstyrene. The inhibitor p-tert-butyl catechol is added in the second step where technical grade DVB is obtained from a column operated without reflux. The yield of DVB distillate relative to the EDB fed to the reactor is about 30%.

The secondary products from this process apart from the DEB and ethyl styrene fraction are benzene, toluene, ethyl benzene, xylene, styrene, methyl styrene and butyl benzene. By adding the inhibitor to the DEB and styrene fraction it may be recycled to the reactor to produce more DVB without too much concern about plugging the reactor with polymer. If this fraction is adequately diluted with DEB the addition of an inhibitor to the reactor becomes unnecessary. In either case this recycle increases the yield and productivity of technical DVB considerably.

4.2 DEHYDROGENATION OF BUTANE, BUTYLENES, ISOPENTANE AND ISOPENTENE

4.2.1 General

The importance of butadiene on the international level has grown at an unprecedented rate in the past 15 years. In the USA from 1955-1970 production doubled reaching about 1.2 million tons per year while in Japan during the same period the production increased about 13 times. The primary demand for this product comes from its use in the manufacture of synthetic elastomers such as cis-1-4-polybutadiene and polybutadiene-styrene.

The raw material used almost exclusively in the industrial production of butadiene is derived from the light fractions end of petroleum refineries. It comprises or consists of the butane-butene-butadiene fraction. The processing of this fraction will be discussed later. A secondary source of butadiene comes from steam cracking of gasoline and is of lesser importance because it occurs as a secondary product in a ratio of 1:2:4 with propylene and ethylene. Further particulars on butane-butylene dehydrogenations may be found in references 1 to 5.

The dehydrogenation of isopentane and isopentene is discussed fully in references (320-327). Because the treatment and processing of C_4 and C_5 fractions are similar from many points of view, the two are treated in this chapter in parallel. The basic chemistry is outlined here for each process. Isoprene used for the manufacture of isobutyl rubber is produced almost entirely by recovery from products formed when gasoline vapours are cracked with steam. The isoprene used in manufacturing 1-4-polyisoprenic rubber is obtained from acetylene, propylene and isobutylene as well as from the process discussed here.

Dehydrogenation Routes to Butadiene

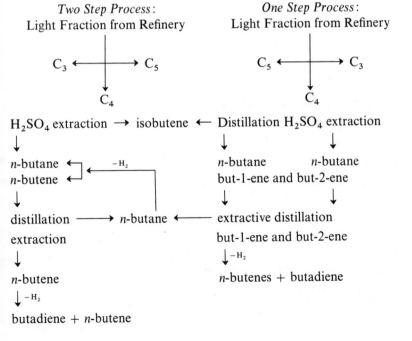

Dehydrogenation of Isopentane and Isopentene

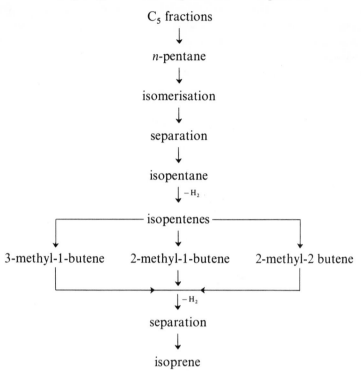

4.2.2 Thermodynamic Aspects, Chemical Kinetics, Reaction Mechanisms and Technological Kinetics

THE CASE OF DEHYDROGENATION OF n-BUTANE AND BUTENES

Reactions and Mechanisms

The dehydrogenation of n-butane yields five major products as shown by the following equations[328,329]

Reaction	ΔH (kcal/mol)
$CH_3CH_2CH_2CH_3 - H_2 \rightarrow CH_2{=}CHCH_2CH_3$	$+31.4$
$CH_3CH_2CH_2CH_3 - H_2 \rightarrow CH_3CH{=}CHCH_3(cis)$	$+28.8$
$CH_3CH_2CH_2CH_3 - H_2 \rightarrow CH_3CH{=}CHCH_3(trans)$	$+27.7$
$CH_3CH_2CH_2CH_3 - H_2 \rightarrow CH_3-CH{=}CH_2 + CH_4$	$+17.6$
$CH_3CH_2CH_2CH_3 - H_2 \rightarrow CH_2{=}CH_2 + CH_3-CH_3$	$+17.0$

The mechanism for these reactions is believed to be of the homolytic radical type.[330] In the initiation step butyl, propyl, ethyl and methyl radicals are formed. In the chain propagation step butenes, propylene and ethylene are formed while the termination reactions produce butane, propane, ethane and methane. The following reactions are representative of the mechanism proposed.

$$C_4H_{10} + H' \rightarrow C_4H_9' + H_2$$
$$C_4H_9' \rightarrow C_4H_8 + H'$$
$$C_4H_9' \rightarrow C_3H_6 + CH_3'$$
$$C_4H_{10} + CH_3' \rightarrow C_4H_9' + CH_4, \text{etc.}$$

The reaction system is complex and a wide range of products is possible from interactions among the various intermediate and secondary products.[331] An example of a series of several probable reactions is given below.

$$C_4H_8 \rightarrow H_2 + CH_4 + C_2H_4 + C_3H_6 + C_4H_6 + C$$
$$\text{(non-catalytic cracking)}$$

$$2C_4H_6 \rightleftharpoons C_8H_{12} \quad \text{(non-catalytic dimerisation)}$$

$$C_4H_6 \rightarrow CH_4 + C_3H_6 + C_2H_6 + C \quad \text{(catalytic cracking)}$$

$$C_4H_8 \rightarrow C_4H_6 + H_2 \quad \text{(catalytic dehydrogenation)}$$

$$C + H_2O \rightarrow CO + CO_2 + H_2 \quad \text{(regeneration of catalyst with steam)}$$

It can be seen that the number and type of reactions places a special demand on the activity and selectivity of the catalyst in particular when high yields and high product purity are specified. One quality which has been stated to be prerequisite for optimum performance in dehydrogenation of mixtures of butane and butene is the capacity of the catalyst to establish an equilibrium between the dehydrogenation reaction and the one for formation of carbon.[332]

Conversion, Yield and Selectivity

Numerous studies have been made aimed at establishing the conditions required for optimum conversion, yield, selectivity, productivity, product purity, etc.[403,404]

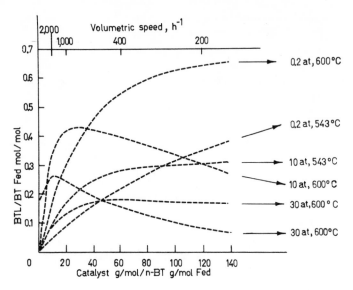

Fig. 43   Effect of catalyst loading on the conversion of butene to butene at different pressures
and temperatures.
BT—butane ; BTL—butene.

The effects of temperature and pressure on the conversion of butane to butenes are shown in Figure 43. The contrary effects of pressure are particularly noticeable. For example at 600°C and a space velocity of $180 \, h^{-1}$ (or a catalyst load of about 140 grams catalyst per mol of *n*-butane feed it can be seen that the mol fractions of butene to butane obtained were 0·65, 0·26 and 0·05 respectively for pressures of 0·2, 10 and 30 atmospheres. This negative effect of pressure on conversion however did not persist at lower loads. Thus, at 600°C the molar ratio achieved at 0·2 and 10·0 atmospheres was the same (0·43) at a load of about 33 grams of catalyst/gmol-*n*-butane feed while at still lower loadings the effect of pressure on conversion was found to be positive. The effect of temperature also was shown to be somewhat dependent upon the load but in the range studied conversions were found to always increase significantly with temperature. Thus at a load of about 90 g catalyst per mole of butane (a space velocity of $300 \, h^{-1}$) the molar ratio of butene to butane obtained at 600°C was 0·62 or more than double that (0·3) obtained at 543°C.

The variations of yield as a function of conversion at different temperatures and pressures is shown in Figure 44. An analysis of the graphs leads to the conclusion that for maximum conversion and maximum yield the

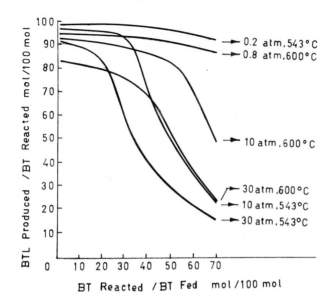

Fig. 44 Variation of yield of butene as a function of conversion of butane at different pressures and temperatures.

BT—butane; BTL—butene.

process should be carried out at low pressures and high temperatures. The temperature, however, should not be so high as to favour the formation of secondary products or to cause excessive thermal decompositions with the ultimate formation of carbonaceous deposits.

The effects of load (grams catalyst per mol butane feed) or space velocity on the conversion and yield values are graphed in Figure 45. At loads under 40 grams of catalyst per mol of butane (and a simultaneous increase of space velocity to values above about 700 h^{-1}) the amount of unreacted butane increases quite rapidly. At higher loads (or lower space velocities) the changes in conversion with changes in loading are small.

Finally in Figures 46 and 47 the equilibrium compositions of the reactor products butane, butene, butadiene and acetylene are shown for various temperatures and pressures. The effect of pressure on the composition of products may be inferred best from Figure 46. The butane content is seen to decrease appreciably with temperature but is also substantially decreased at all temperatures by a reduction of pressure from 1 atmosphere to 125 mm Hg pressure. Thus at 300°C the mole fraction of butane is reduced from 1 to 0·15 for this change in pressure, and at 700°C from 0·1

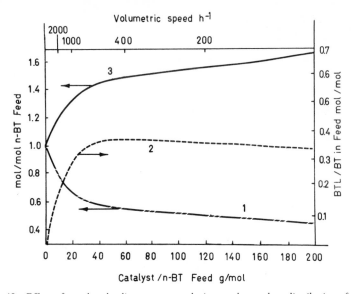

Fig. 45   *Effect of catalyst loading or space velocity on the product distribution of butene dehydrogenation.*
Graph 1—unreacted n-butane; graph 2—butene formed; graph 3—total products formed.

to 0. The content of butene however is altered more selectively by temperature and pressure. At 300-500°C the butene mol fraction ranges between 0·73 and 0·68 but decreases sharply as the temperature is increased further at 125 mm Hg pressure. At 1 atmosphere pressure on the other hand, the butene content begins to increase at about 350°C and reaches a maximum mol fraction of about 0·65 at about 670°C. (At 125 mm Hg pressure the butene mol fraction at this temperature is only about 0·33).

The butadiene content is somewhat higher at lower temperatures and lower pressures. For example at 125 mm Hg pressure a mol fraction of about 0·9 is reached at about 800°C whereas at 1 atmosphere pressure even at 900°C the mol fraction obtained is only about 0·8.

The graphs indicate also that at temperatures above 800°C the content of acetylene increases quite rapidly at both pressures, the rate of increase being somewhat more rapid at the lower pressure.

In Figure 47 the conversions of butane to butene, butadiene and acetylene are graphed as function of pressure at 530 and 593°C. This form of presentation shows more clearly the direction in which the temperature and/or pressure should be altered to favour the production of butenes or of butadiene. At these temperatures the production of acetylene is negligible

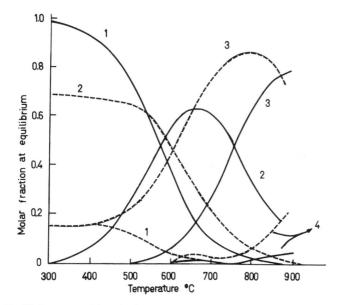

Fig. 46 Equilibrium compositions for butane dehydrogenation reactions as a function of temperature and pressure.

Solid line graphs indicate mole fractions of equilibrium mixture components at 530°C and broken line at 593°C. Graph 1—butane; graph 2—total butenes; graph 3—total butadienes; graph 4—total acetylenes.

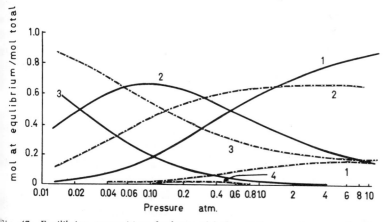

Fig. 47 Equilibrium compositions for butene dehydrogenation reactions as a function of temperature and pressure.

Solid line graphs indicate mole fractions of equilibrium mixture components at 530°C and broken line at 593°C. Graph 1—butane; graph 2—total butenes; graph 3—total butadienes; graph 4—total acetylenes.

at all pressures. The yield of butadiene on the other hand increases sharply with increases in temperature and reductions in pressure. Total conversion of butane apparently can be effected at pressures of about 0·01 atmospheres. This is still only of academic interest since at present industrial scale operations under 0·2 atmospheres are impracticable. Thus it can be concluded that a yield of about 0·42 mols of butadiene per mol of total product may be considered as near the maximum attainable at 593°C.

Oxidative Dehydrogenation

The equilibrium reactions of dehydrogenation of butane or butenes can be displaced to the right by fixing the hydrogen formed with oxygen or iodine,[338-340] chlorine, bromine, sulphur[341] or with SO_2[405,406] as typified by the following reactions:

$$C_4H_8 + \tfrac{1}{2}O_2 \rightarrow C_4H_6 + H_2O$$

$$C_4H_{10} + 2I_2 \rightarrow C_4H_6 + 4HI$$

$$C_4H_8 + Cl_2 \rightarrow C_4H_6 + 2HCl$$

Oxidative dehydrogenation with oxygen (air): in the case of producing butadiene by dehydrogenation of butenes in the presence of air the following clarifications can be made.

The yield of butadiene increases with temperature and with increased ratio of air to butene in the feed. For instance at an air to butene ratio of 7·2:1 the yield of butadiene increases from about 15% at 300°C to about 45% at 400°C. When the air to butene ratio is reduced to 6:1 the yield at the lower temperature is unaffected but drops to about 43% at the higher level. Similarly at a ratio of 4·7:1 the yield at 400°C is affected more strongly falling off quite sharply to about 35%.

The conversion of butene also increases with temperature. At an air to butene ratio of 6:1 the conversion increases from about 20% at 300°C to about 60% at 400°C. The yield of butadiene relative to the butene reacted can thus be seen to remain relatively constant (about 70-80%) for air to butene ratios between 4·7 and 7·2 to 1 in the temperature interval 300 to 400°C. To effect higher yields of butadiene using air as the oxidative dehydrogenation agent it is evident that for reactions at 400°C the minimum air to butene ratio should be of the order of 6-8 to 1. (It is very probable that this range also encompasses the maximum.)

The dehydrogenation of butenes to butadiene also can be carried out with oxygen. For example at 500°C and a butene-oxygen ratio of 1:0·5 the yield in butadiene was about 50 % and the selectivity was better than 90 %. At lower ratios for the same catalyst the yield of butadiene increased while the selectivity decreased slightly.

The dehydrogenation of butenes (1-butene, *cis*-2-butene and *trans*-2-butene) over catalysts based on oxides of tin and antimony (Sn : Sb atomic ratio of 1 : 4) in the presence of oxygen, also, has been studied.[333] Under continuous operation the conversion levels found were in the ratio of 1·0 : 0·46 : 0·51 for 1-butene; *cis*-2-butene; *trans*-2-butene to butadiene. The reactant concentration was found to have a minor influence on the degree of conversion. For example an increase in the concentration of 1-butene from 0·9 to 1·38 % was found to decrease the conversion to butadiene from 82 to 81 %. The oxygen concentration, however, was found to have little influence on the degree of conversion. In comparison with catalysts based on oxides of bismuth and molybdenum for the oxidative dehydrogenation of 2-butenes to butadiene, the activity of the former was noted to be vastly superior (0·54 versus 0·23).

Oxidative dehydrogenation with iodine: commercial oxidative dehydrogenation processes based on iodine are feasible only if economic (i.e. virtually complete) recovery of the iodine (and hydrogen iodide formed upon reaction) is effected. Although this can be achieved in several ways the most attractive method appears to be one which employs metal oxides which react with the iodine bearing vapours to form metal iodides from which the iodine is subsequently recovered by reaction with warm air with the attendant regeneration of the metal oxides. Copper oxide and mixed oxides of manganese and potassium on a support have been recommended as acceptors.

The conversion of butane at 520°C and a space velocity of 460 h^{-1} was found to increase from about 22 % to 100 % when the molar ratio of iodine to butane was increased from 0·2 to 1·4. The yield of butene under the same conditions increased from about 10 to about 40 % and that of butadiene from about 5 to 28 %. The selectivity for the formation of butene and butadiene was found to vary from 80 to 96 % with the maximum selectivity occurring at a molar ratio of 0·8-1·0 for iodine-butane. The yield in butene relative to the reacted butane was found to increase with increases in space velocity between the values of 260 h^{-1} and 500 h^{-1}. The amount of butane converted per pass over this variation of feed rate was negligible.

At 540°C, a space velocity of 500-520 h^{-1} and molar ratio of iodine to butane of 0·6-0·7 to 1, dilution of the reactants with inert gas (nitrogen) was found to reduce secondary reactions considerably. Thus, for instance an alteration of the nitrogen to butane ratio from 1:1 to 2:1 increased the yield of butadiene from 40 to 45% and the selectivity for butene and butadiene from about 68 to about 78%.

Another process which has been applied to the dehydrogenation of *n*-butane recovers iodine and hydrogen iodide by reaction with ammonia.[334] The following reactions have been proposed:

$$3I_2 + 2NH_3 \rightarrow N_2 + 6HI$$

$$HI + NH_3 \rightarrow NH_4I$$

Studies at atmospheric pressure in a continuous flow system showed that the optimum temperature limits for fixation of HI as NH_4I were 200-250°C for a molar ratio of NH_3 : HI of 4:1 and a reactor retention time of 3 seconds.

Catalysts: a number of high performance catalysts used commercially for dehydrogenation of and butene are compared in Table 18.[319] The performance of the above catalysts (and many others) has been studied with respect to various kinetic and technological considerations. Thus as an example the Dow catalyst based on Cr_2O_3 promoted $Ca_3Ni(PO_4)_6$ has been studied in a fixed-bed continuous flow reactor to determine the effect of compositional changes on its activity for dehydrogenation of butenes.[335] The gas compositions were determined by gas chromatography and the catalyst characteristics defined by X-ray analysis and by the application of thermic methods. The following observations and conclusions were made.

At 570° for Ca-Ni phosphate and 670°C for Ca-Cr-phosphate distinct exothermic effects were noted. Since the samples were found by X-ray analysis to be amorphous at lower temperatures these exothermic effects were attributed to crystallisation of the phosphates at the higher temperatures. Similarly the endothermic effects noted at lower temperatures were attributed to the decomposition of crystalline hydrates. The ternary Cr-Ca-Ni phosphate also displayed endothermic effects at low temperatures (260-340°C) and exothermic effects at high temperatures (690-900°C). Again the X-ray diffraction patterns revealed that these samples were constituted of solid solutions similar to that of the Ca-Cr phosphates. Thus it was concluded that this catalyst is comprised of a mixture of tertiary and binary solutions with small amounts of Ca-Cr phosphate and Cr_2O_3.

TABLE 18

Characteristics of butane-butene dehydrogenation catalysts

Commercial spec.	Composition (%)	Temperature (°C)	Space velocity (h^{-1})	Conversion (%)	Selectivity (%)	Regen. agent	Reference
1707 (butene)	18·4 Fe_2O_3 4·6 K_2O 72·4 MgO 4·6 CuO	630	400	20·2	70-73	—	336
105 butene	90 Fe_2O_3 4·0 Cr_2O_3 6·0 K_2CO_3	620	400	19-75	69-75	Steam	336
Dow (butene)	$Ca_3Ni(PO_4)_6$ +2% Cr_2O_3 on support	565-663	125-175	35-45	90	Air and steam	337
Houdry (butane) (butene)	$\left.\begin{array}{l}Al_2O_3\\Cr_2O_3\end{array}\right\}$ 18-20	593-677 600-621		42·5-49·5	90-94 62	— air	338

The catalyst activity was studied at 600°C and 650°C with respect to butene dehydrogenation. Of the individual components Ni phosphate was found to be most active and of the binary Ca-Ni phosphate. The activity of Ca-Cr phosphate was negligible. Hence it was concluded that nickel ions are the effective active component of this catalyst. The Ca phosphate serves as a support to fix them in a solid solution and the Cr_2O_3 acts to enhance their activity at the surface.

Although the results obtained from differential reactors differ from and sometimes contradict those obtained from integral reactors it is generally agreed that the former provide more conclusive data on the technological performance of catalysts.[339-342] Some of the conclusions reached about the butane to butene dehydrogenation reaction from studies carried out in a differential microreactor at 480-520°C, with a maximum conversion of 5 mole %, are given below. The feed rate of n-butane and nitrogen, in these studies, was varied from 15 to 36 standard litres per hour and the partial pressure of n-butane in the feed from 0·45 to 1·13 atmospheres. The catalyst bed contained 2 grams of catalyst of the $Cr_2O_3.Al_2O_3$ type.[343]

Analyses of the data indicated that the rate determining step for this dehydrogenation reaction on this catalyst is the surface reaction. The rate expression deduced, based on the assumption that the Langmuir-Hinshelwood 'two centre' type of mechanism applies and that the reverse reaction is negligible,[342] was as follows:

$$(P_B/n)^{\frac{1}{2}} = \frac{1}{(k_r b_B)^{\frac{1}{2}}} + \left(\frac{b_B}{k_r}\right)^{\frac{1}{2}} P_B$$

where

k_r = the reaction rate constant

b_B = the absorption equilibrium constant

P_B = the partial pressure of n-butane

The dependence of log k_r on the reaction temperature was found to be of the classic Arrhenius type. This enables the researchers readily to deduce values for the activation energy and the frequency factor for the k_r expression.

Compounds of bismuth and molybdenum also have been found to have high activities and selectivities for the oxidative dehydrogenation of olefins. Additions of small quantities of iron oxide to these catalysts have been found to increase their activities substantially. The main particulars and conclusions of an extensive study of the promoter effect of Fe_2O_3 on the $Bi_2O_3MoO_3$ type of catalysts[344] are given below.

In this study the activities of the individual and mixed components of this catalyst system were defined and determined as the average rates of formation of butadiene, carbon dioxide and butene-2 from butene-1 in the temperature range 300-550°C. The interactions between the components of the various catalytic systems were determined with the aid of differential thermograms and X-ray diffraction pattern studies of the individual oxides and of mechanical mixtures of these oxides. The specific surfaces of the various components and mixtures also were determined. Specific catalysts were prepared by depositing the material on ceramic granules 1·5-2·0 mm in size which were assumed to have a uniform specific surface of 0·4 m²/g. The specific surface areas of other materials used were as follows:

Material	Atomic ratio	Specific surface (m²/g)
Bi_2O_3		0·7
Bi:Fe	1:1	4·2
Mo:Bi	3:2	1·5
Mo:Bi:Fe	3:1:1	1·6
Mo:Fe	3:2	3·3
MoO_3		1·6
Mo:Bi	2:2	1·9
Mo:Bi:Fe	2:1:1	1·5
MoFe	2:2	4·2
Fe_2O_3		9·2

The pertinent conclusions of this study were as follows. The ternary system contains $Bi_2(MoO_4)_3$ and no free iron oxides. The oxides of iron and molybdenum are active for all reactions (i.e. oxidative dehydrogenation, total oxidation and isomerisation). MoO_3 has a greater activity for isomerisation and Fe_2O_3 favours oxidation. The activation energies for oxidative dehydrogenation total oxidation and isomerisation to butene-2 were found to be 18, 15 and 8 kcal/mol respectively for MoO_3 (at 450°C) and 23, 18 and 9 kcal/mol respectively for Fe_2O_3 (at 350°C). The average rates of formation of butadiene, carbon dioxide and butene-2 were found to be 0·1, 0·1 and 0·6 moles/m² h for MoO_3 at 450°C and 0·1, 0·2 and 0·1 moles/m² h for Fe_2O_3 at 350°C, respectively.

The appreciable increase in the specific activity of MoO_3 when deposited on a ceramic support, relative to unsupported MoO_3, was attributed to not only the increased specific surface of the oxide (as effected by deposition and dispersion throughout the support) but also to the special role that the support plays in the reaction mechanisms. The BiMo catalyst has a higher activity for total oxidation and isomerisation in the absence of a support. On the other hand the ternary Fe-Bi-Mo catalyst is much more active for dehydrogenation reactions than the single molybdates of Bi or Fe. For the ternary systems, maximum dehydrogenation activity was found for an Mo : Bi : Fe ratio of 6 : 3 : 1 up to 7 : 2 : 1 depending upon the temperature.

The addition of Fe_2O_3 to BiMo catalysts enhanced both the selectivity and activity for the formation of dienes. This was attributed to the weakening of the bonds between the catalyst and adsorbed oxygen. The fact that adsorbed oxygen affects the electrical conductivities of these catalysts in agreement with this hypothesis was cited as proof for it.

In the Houdry process for the dehydrogenation of butane to butylene in a single step the thermal balance must be maintained within satisfactory limits. This may be achieved by operating the adiabatic reactor in a cyclic manner to utilise the heat of combustion of the cox deposits as part of the heat requirement. This requires that the heat developed in the regeneration period of the catalyst be stored for use during the reaction period. To accomplish this with optimum heat economy an inert material is generally admixed with the catalyst. Since the prime purpose of this inert material is to accumulate the heat generated during the regeneration cycle and to release it as high an inert to catalyst ratio as acceptable from the reactivity consideration during the reaction cycle may be used. Fire brick, kaolin, quartz and fused aluminum oxide are inerts which have been utilised in this way. The inert to catalyst ratio by weight required to effect a satisfactory balance between thermal economy and reactivity requirements has been found to be about 3 to 1.

The use of aluminium oxide as an inert heat carrier in the vapour phase oxidation of n-butane, isobutane, 1-butene, 2-butene and isobutylene, also has been studied intensively at 350-550°C and pressures of 0 to 100 psig.[341] The fused alumina was introduced into the reactor as 'rain', counter-current to the gas flow. Two sizes (35-40 mesh and 80-100 mesh) of alumina were used. These occupied less than 1% of the reactor volume at any instant. The ranges of independent variables studied were as follows:

	n-*butane*	*isobutane*	*2-butene*	*1-butene*	*isobutylene*
Temperature (°C)	350-560	400-560	350	380	350-560
Molar ratio O_2/HC	0·50, 0·25	0·50	0·25	0·25	0·25
Pressure (psig)	0, 50	0	50, 100	50, 100	50, 100
Oxidant	O_2, air	O_2	air	air	air
Solids rate (lbs/ gm mole O_2)	0-28	0-27	0-37	0-20	0-34

The following observations and conclusions were made. In the case of oxidation of *n*-butane and isobutane at 450°C and 550°C at atmospheric pressure the introduction of particulate solids greatly increased the selectivity to butylenes but reduced substantially the yield of propylene, ethylene and paraffins. When more and more particulate surface was employed at 550°C this situation was reversed. This suggests that the presence of solids favoured hydrogen abstraction reactions initially rather than subsequent disproportionation and cracking reactions, however, when excess surface was present the latter became dominant. It also was noted that the yields of oxygenated compounds increased with increasing solids flow rates when the olefins were oxidised in this manner. This suggests that olefins may be selectively oxidised in olefin-paraffin mixtures in the presence of particulate solids.

DEHYDROGENATION OF ISOPENTANE AND ISOPENTENE

The equilibrium compositions of isopentane, isopentene and isoprene at 0·2 and 1·0 atmospheres pressure are graphed as a function of temperature in Figure 48.[346] If can be seen that the formation of isoprene is favoured by low pressure and high temperatures.

From these graphs and those in Figure 49 which represent the equilibrium compositions as attained by dehydrogenation of a mixture of isopentane and isopentene it is evident that conversion to isoprene is favoured also by higher concentration levels in the feed.

An important aspect of isopentane dehydrogenation is that the yields of isopentene and isoprene are dependent not only upon the rates of the dehydrogenation reactions and cox formation, but also upon the rate of isomerisation which leads to the formation of *n*-pentane, *n*-pentene and

1-3-pentadiene. The kinetic aspects of these isomerisation-dehydrogenation reactions have been characterised in a semi-industrial scale fluid-phase reactor.[347] The reaction temperature was 540-580°C in the upper part of the catalyst beds and 30-40°C less in the lower layers.

It was found that a sectioned reactor containing 12 catalyst beds yielded 5% more isopentene than one containing 8 beds. The catalyst beds in both instances were pretreated by passing butane at 640-650°C at a volumetric space velocity of 100 h⁻¹. In the 12 bed reactor the yield decreased by about 10% based on the isopentane used when the temperature was raised from 540 to 580°C. The amount of pent-1-ene and pent-2-ene (in a ratio of 1:3·5) and of pentadiene formed was found to increase with increased yields of isopentene by pretreatment of the catalyst. The yield of pentadiene, for example was found to be 20-60% greater when a treated rather than an untreated catalyst bed was employed.

These data are summarised in Figure 50. With regard to *n*-pentane it was noted that about 1·5-2 times as much was formed when the catalyst was preheated with butane than when it was not.

The rate of isomerisation to *n*-pentane however was found to decrease rapidly as the percentage of this component in the feed increased. Indeed at about 7·5% the isomerisation activity was reduced practically to zero.

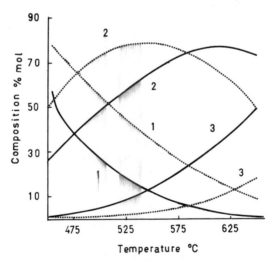

Fig. 48 *Equilibrium hydrocarbon compositions as a function of temperature for isopentane dehydrogenation reactions.*
Graph 1—isopentane; graph 2—isopentene; graph 3—isoprene; − − − − *at 1·0 atmosphere pressure;*   *at 0·2 atmosphere pressure.*

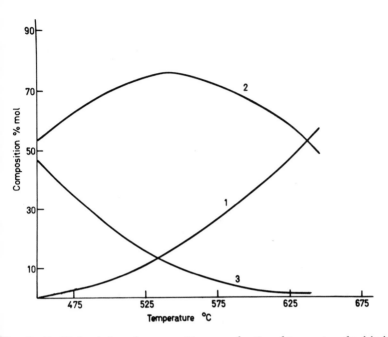

Fig. 49 Equilibrium hydrocarbon compositions as a function of temperature for dehydrogenation of a mixture of isopentane and isopentene.
Graph 1—isopentane; graph 2—isopentene; graph 3—isoprene; all at 0·2 atmosphere pressure.

Thus to improve the yield of isopentene and isoprene and to reduce the amount of isomerisation of isopentane to n-pentane the catalyst should be pretreated with residual gases which contain hydrocarbons and a feed stock should be used which contains about 7 % (by weight) of n-pentane.

Kinetic studies carried out on the dehydrogenation of the three methyl butylenes, taken individually have shown that 2-methyl-2-butylene dehydrogenates most rapidly.[348] The kinetic rate constants expressed in $(cm^3 min/cm^3)$ at 560, 580, 590 and 600°C for 2-methyl-2-butylene, 2-methyl-1-butylene and 3-methyl-1-butylene were determined to be respectively 7·39, 5·92 and 5·13; 9·39, 8·32 and 6·90; 10·93, 9·36 and 7·99; and 13·29, 10·93 and 9·03.

The foregoing data indicate that in the case of a single step dehydrogenation of isopentane over specific catalysts increased yields of isoprene will be obtained at higher temperatures. That is the equilibrium, mono-olefin ⇌ di-olefin, shifts to the right as the temperature is raised. On the other hand as the pressure is lowered increased conversion to isoprene

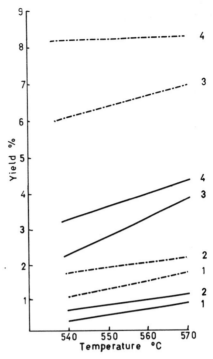

Fig. 50   *Effect of temperature on the field of* n-*pentene and pentadiene during dehydrogenation of isopentane.*
Graphs 1 and 2—yield of pentadiene; graphs 3 and 4—yield of n-*pentene; graphs 1 and 3 with untreated catalyst; graphs 2 and 4 with butane treated catalyst.* —— *based on isopentane circulated;* ———— *based on isopentane reacted.*

also, is effected. However, in this case the reason differs. At higher pressures higher conversions of isopentene occur but yields of isoprene are reduced due to more extensive formation of cox. It also should be noted that higher space velocities lead to lower yields of isoprene.

The oxidative dehydrogenations of isopentene using oxygen (air) and/or iodine also have been reported.[349-354] Thus according to these references when a mixture comprised of 3·31 % n-pentane, 9·9 % n-pentene, 5·1 % isopentane and 80 % isopentene was reacted at 250°C at a space velocity of 0·11 h^{-1} in the absence of oxygen but at a molar ratio of iodine : water vapour; isopentene of 0·05:8:1 the isopentene conversion, yield of isoprene and selectivity of reaction in this regard were found to be 20, 8 and 40 % respectively. With the introduction of oxygen in a molar ratio with respect to isopentene varying from 0 to 1·5 the conversion, yield and selectivity were found to increase significantly. For example at unity molar ratio of oxygen to isopentene the selectivity was about 93%.

For the same mixture and reaction temperature but with an iodine: oxygen:water vapour molar ratio of $0.05:1:12.4$ the conversion, yield and selectivity also were found to increase with increased contact time (i.e. reduced space velocities). Thus at space velocities below about $0.13-0.2\ h^{-1}$ yields and selectivities of isoprene formation up to about 75 and 94.5% respectively were achieved whereas at higher space velocities these values diminished appreciably. At a space velocity of $0.11\ h^{-1}$ and a molar ratio of iodine:oxygen:water vapour:isopentene of $0.05:1:12.4:1$ it was noted that an increase of the reaction temperature from 450°C to 500°C increased the conversion, yield and selectivity values. Above 500°C these values remained constant or decreased.

The dehydrogenation of isopentane to isoprene in the presence of iodine, oxygen and water vapour proceeds via the reactions:

(a) $i\text{-}C_5H_{12} + I_2 \rightleftharpoons i\text{-}C_5H_{10} + 2HI$

(b) $i\text{-}C_5H_{10} + I_2 \rightleftharpoons i\text{-}C_5H_8 + 2HI$

(c) $4HI + O_2 \rightleftharpoons 2H_2O + 2I_2$

Reaction (b) the dehydrogenation of isopentene proceeds with greater speed and selectivity than reaction (a), the dehydrogenation of isopentane. Hence for this reason as well as that of a lower iodine consumption the feed stock should have a high content of isopentene. The dehydrogenation rate constant of isopentene is more than three times that of isopentane under the selected conditions referred to earlier. Furthermore an additional benefit accrues from using feed stocks rich in isopentene in that losses to complete oxidation (to CO and CO_2) and decomposition of product are reduced. These were found to be 1.2% and 1.4% respectively for isopentene dehydrogenated with iodine as compared with 3.8% and 4.7% respectively for dehydrogenation of isopentane with iodine.

A catalyst typical of the type employed in oxidative dehydrogenation of butylenes and isopentenes is that based on $SnO_2\text{-}Sb_2O_4$. This catalyst system has been studied over wide temperature ranges under continuous and discontinuous reaction conditions. The following pertinent conclusions were reached[355]

(a)   Technical grade antimony oxide does not catalyse the oxidation reactions.

(b)   The yield of butadiene or isoprene increases with increasing ratio of Sn:Sb, reaching a maximum at 4:1. At 450°C the yield obtained of butadiene was 70% and of isoprene, 17%. The yield of CO_2 also increases with increasing content of tin.

(c) The catalytic activity for the isomerisation reactions of olefins reaches a maximum at higher contents of antimony than does that for the oxidation reactions.

(d) The surface properties of this catalyst system also vary with the composition. The largest specific surface was found for a Sn:Sb ratio of 4:1 which corresponded to that for the greatest activity for diene formation.

(e) The individual components are less active than the various combined mixtures of the two oxides in catalysing the selective oxidation of olefins. The mixed oxides react to form amorphous materials with large specific surfaces and high activities as evidence by X-ray analysis and reactivity tests, respectively.

Catalysts of the chromium-aluminium oxide and chromium-oxide platinum type also have been tested and found effective for the production of isoprene by oxidative dehydrogenation.[356] It was found that additions of rare earth oxides of the lanthanide series had a small influence on the activity and selectivity of the chrom-alumina catalyst system whereas additions of alkaline earth oxides improved both the activity and selectivity considerably. The chrom-platinum catalyst system was tested at 530°C and a space velocity of 1.2-2.0 h^{-1}. The rare and alkaline earth addition to it were made by treating the bulk catalyst with 10% nitrate solutions of cesium, rubidium, cerium, ytterbium and niobium.

4.2.3 Technological Aspects

THE CASE OF BUTANE AND/OR BUTYLENE DEHYDROGENATION

Processes Other than the Houdry Process

The main characteristics of several processes other than the Houdry process for the dehydrogenation of butane and/or butylene are given in Table 19. The specific differences lie in the way that thermal equilibrium is maintained or realised in the reactor and in the way which contact is established between the catalyst and the reactant mixture as well as in the physical and chemical characteristics of the reactor. Particulars about the Houdry process are given in the next section.

The fixed bed dehydrogenation process with external heating is shown

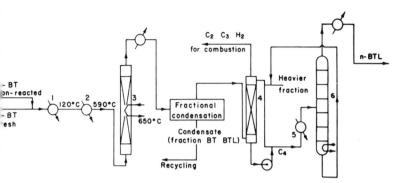

Fig. 51 Continuous process for manufacture of butylenes by dehydrogenation of n-butane in a fixed bed reactor with external heating.
—preheater at 120°C; 2—preheater at 590°C; 3—externally heated reactor; 4—scrubber (absorber); 5—heater; 6—desorption column; BT—butane; BTL—butylenes.

chematically in Figure 51. It can be seen that the reactant mixture is preheated successively by two preheaters prior to introduction into the reactor. The effluent is cooled and fractionally condensed by further cooling and compression in alternate steps. The resulting condensate (comprised mostly of butane but containing some butylenes) is recycled, whereas the remaining gases are scrubbed in the absorber 4 to yield light gases, C_2, C_3 and H_2, which are drawn off to the combustible gases stream and the butylene extract which is recovered and further purified in the desorption column 6.

Dehydrogenation of *n*-butane without external heating and using a mobile spherical catalyst requires two reactors, one of which is employed continuously to regenerate (and preheat) the catalyst. Adequate heating is provided by burning the combustible gases referred to above. The butylene products obtained from this reactor are close to the equilibrium composition.

The fluidised bed process is shown schematically in Figure 52, where it can be noted that the butane feed is first preheated by the reactor gases and then contacted with the catalyst in a fluidised state within the reactor. Inert gases are fed to the reactor to control the degree of fluidisation and temperature in the bed and regenerated (and fresh) catalyst to maintain constant bed activity. The catalyst is circulated pneumatically and is

TABLE 19

Process parameters for six catalytic processes for dehydrogenation of *n*-butane (BT) and/or *n*-butylene (BTL)

Process number / Process parameter	1	2	3	4	5	6
Feed composition	BT	BT	BT	(a) C$_4$ fraction 26-38·5 mol % 1BT 36-48 mol % BTL 6·5-9·5 mol % BT Other products or (b) BT-BTL fraction 78 wt % BTL 15 wt % BT other products	—	—
Temperature (°C) at entrance to reactor	590	610-620	550	650	—	—
Reaction temperature (°C)	650	575-600	580	580-650	520-550	520-550
Temperature (°C) at exit to reactor	—	—	—	580-590	—	—
Contact time (seconds)	1·8	—	—	—	—	—
Reactor type	Tubular, vertical	Vertical pipes	Fluidised bed	Vertical adiabatic	Adiabatic	Adiabatic
Diameter/length (m)	0·05/3	—	—	—	—	—

	1	2	3	4	5	6
Catalyst		H_2O / Spherical particles circulate with reactant		CuO 4.6%+ or $Ca_3Ni(PO_4)_6$ Cr_2O_3 2% or Fe_2O_3 90%+ Cr_2O_3 4%+ K_2CO_3 6%	$Bi:M_0 = 0.7\text{-}0.9$	transition metal halogen on Al_2O_3 or SiO_2 support
Reactor products (exit composition (%))	BT, BTL, C_3, C_2 etc.	BT, BTL, etc.	BT 52.3 Vol%+ BTL 31.1 Vol%+ C_3 5.3 Vol%+ C_2 3.8 Vol%+ Other products	BT 3.3 mol% BTL 51.2 mol% BTD 33.5 mol% C_4 9.1 mol% C_3 1.3 mol% Other products	—	—
Space velocity	2100 1/1/h	200 m³/m³ h (200°C)	120 m³/m³ h	125-175 m³/m³ h	200-400 h⁻¹	430 h⁻¹
Conversion	25%/pass 80% with recycle	20%	33%	45-55 mol%	—	80-81%
Selectivity	—	80-85%	70%	90-94 mol%	70-90%	75-78%
Yield	—	16-17%	74% BTL and BTD relative to reactor feed	—	50%	43-46% in 1-3 BTD 15-19% in BTL
References	—	357	358	336, 337	359, 360	361, 363

Process Definitions

1 Catalytic dehydrogenation of *n*-butane to *n*-butylene in fixed bed reactor with external heating.
2 As for (1) but with mobile catalyst and without external heating.
3 As for (1) but using a fluidised bed reactor.
4 Catalytic dehydrogenation of butylene to butadiene.
5 Oxidative catalytic dehydrogenation of *n*-butane and *n*-butylene.
6 Catalytic dehydrogenation of *n*-butane and *n*-butylene in the presence of iodine.
BT—butene; IBT—isobutene; BTL—butene; BTD—butadiene.

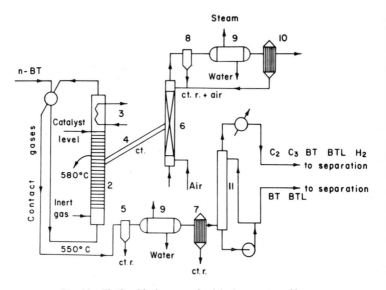

Fig. 52　Fluidised bed process for dehydrogenation of butane.
1—heat exchanger for preheating n-BT to 550°C; 2—fluidiser bed reactor; 3—cooling coil; 4—regenerated catalyst return pipe; 5,8—cyclones; 9—recuperators; 7,10—electrostatic filters; 11—absorber, scrubber; ct.r.—recovered catalyst; BT—butane; BTL—butene.

recovered with high efficiency from the reactor effluent by the series of cyclones 5, recuperator 6 and electrostatic precipitator 7 and from the regenerator by the cyclone 8, recuperator 9, and electrostatic precipitator 10 circuit.

The reactor effluent gases are pre-cooled by the cooling coil 3 situated just above the fluid bed interface to quench secondary reactions, thereby improving the yield and reducing the purification problems. On exit from the electrostatic filter 7, these gases are scrubbed in the absorption column 11 to produce a butane and butylene fraction and a C_2, C_3 and H_2 fraction which also contains some butane and butylene. Both the extract fraction and the non-absorbed gases are sent to further separation processes.

Dehydrogenation of C_4 olefins to butadiene is carried out adiabatically by injecting steam into the reacting mixture of butylenes to maintain the temperature within practical limits and to reduce the rate of catalyst deactivation due to carbon deposition. The steam is introduced or generated primarily from the water spray which is utilised to quench the reactor effluent.

In the oxidative dehydrogenation process using air, thermal balance is achieved by regulating the temperature and the degree of butylene dilution with air to maintain a balance between the rates of oxidative dehydrogenation and secondary reactions. There is a strong tendency for superheated zones to appear due to non-uniformity of temperature and composition of the reactant gases as well as non-uniformity of flow through the catalyst bed.

The oxidative dehydrogenation process using iodine can be developed optimally only with a mobile catalyst type reactor system. Only in this way could a rapid and complete combination of the hydroiodic acid be ensured; that is by rapid replacement of exhausted with fresh acceptor. The yield of butadiene decreases sharply with contact time, falling from about 70 % at 4 seconds to about 10 % at 20 seconds. A further requirement for the process would be to utilise the heat released in the oxidation of the hydroiodic acid.

The Houdry Process

This process appears to be the most competitive at the present time. The flow diagram for the single step Houdry process is shown in Figure 53

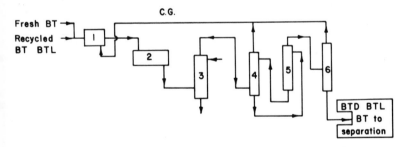

Fig. 53   The single-step Houdry process.
1—furnace; 2—reactor; 3—column for cooling with oil spray; 4—adsorption column; 5—desorption column; 6—column for separation of C_3 fraction; BT—butane; IBT—isobutane; BTL—butene; BTD—butadiene.

and pertinent details about four variations of it are given in Table 20. The Houdry process for producing butadiene by dehydrogenation of

TABLE 20

Pertinent details on four variations of the Houdry process

Process number	1	2	3	4
Process parameter feed composition (%) (total fresh and recycled)	BT-64·3, BTL-24·7 plus other products	BT-95 and other products	BT-95 and other products	BT-30·2, BTL-57·7 and other products
Reaction temperature (°C)	565–620	—	—	—
Reaction pressure (mm Hg)	130–180	—	—	—
Reactor type	Adiabatic, horizontal	—	—	—
Diameter/length (metres)	4·5/10	—	—	—
Product composition at reactor exit (%) (excluding miscellaneous products)	BT 44·7 BTL-25 BTD-11·8 C_4 10·5 C_3 2·1	BT 48·6 BTL 28·5 BTD 6·4 C_4 5·2 C_3 2·8		BT 26·9 BTL 57·7 BTD 14·3 C_4 10·9 C_3 0·6
Catalyst type	Fused alumina, kaolin or bentonite used as inert additive	—	—	—
Space velocity	1·8–2 h^{-1} (liquid)	—	—	—
Selectivity	BTD from BT reacted 57·6 wt %, 61·9 mol %	BTL and BTD from BT reacted 75·3 wt %, 78·4 mol %	BTL from BT reacted 93·5 wt %, 76·1 mol %	BTD from BT and BTL reacted 76·2 wt %, 79·6 mol %
References	338	364	365	366

Process Definitions

1 The Houdry process for dehydrogenation of n-butane to butadiene in a single step.
2 The Houdry process for dehydrogenation of n-butane to butylenes and butadiene.
3 The Houdry process for optimum yield of butylenes by dehydrogenation of n-butane.

butane in a single step has the greatest flexibility in so far as butylenes and butadiene can be obtained in any desired ratio. If butadiene is required as the main product the butylenes are totally recycled with the unreacted butane. The reactors are operated adiabatically and cyclically to regenerate the catalyst by burning off the cox deposits and to recuperate this heat release.

As stated under catalyst type in Table 20, inert additives are employed along with the usual catalyst systems to accumulate heat during the regeneration cycle and release it adiabatically during the reaction cycle. Referring to Figure 53 it can be seen that the fresh butane and recycled butane and butylene are preheated in the furnace 1 to the reaction temperature by the heat released from the combustible gases (C_3, C_2, etc.) prior to introduction into the reactor 2. The reactor effluent gases are cooled in column 3 by an aromatic oil spray to about 45°C. Then, they are compressed and scrubbed in the absorption column 4. The absorbed material is passed to column 5 where the hydrocarbons are desorbed. Propane is separated in column 6 from the butylene and butadiene products and the unreacted butane. After the final separation the mixture of butylenes and unreacted butane is recycled to the feed stream.

The contact mass (catalyst and inert material) is supported on a perforated grid forming a layer about one metre deep. The grid and reactor lining are made of a high chromium ($\sim 25\%$) alloy steel although magnesia and other ceramic materials also are used to line the reactor.

The other variations of the Houdry process as stated earlier and outlined in Table 20 differ with respect to the product objective and/or feed composition employed and conversion and selectivity achieved. The Catadien process developed by Houdry Process and Chemical Co. offers two alternatives. In the first case with moderate conversion 1·89 kg butane are required per kg of butadiene, whereas in the second with high conversion 1·55 kg butane suffice. The main improvements in this process lie in catalyst activity, and manufacturing control which lead to lower production costs. Clearly the capital investment for equivalent butadiene production capacity will be greater for the moderate than the high conversion process since more recyle requires larger installations. However the choice between these alternatives also will be determined by the local cost of butane.

Irrespective of the process the amount of cox deposits produced is about 2·7 kg per 100 kg of butane reacted. As the deposits build up on the catalyst surface, higher and higher temperatures are required to maintain a constant reactivity. Generally the catalyst is regenerated when a temperature of ~ 560°C is required to effect the desired conversions per pass.

SEPARATION AND PURIFICATION PROCESSES FOR BUTADIENE

Irrespective of the manufacturing origin of the butadiene, processes for its separation and purification to be competitive must minimise capital investment and utility requirements.[367]

In the case of butadiene production by dehydrogenation of BT and/or BTL the separation–purification processes are as follows. The separation of the BT–BTL fraction from heavier fractions (C_5 and above) and from lighter fractions (methane, ethane, ethylene, propane, carbon dioxide, hydrogen, etc.) may be effected by partial condensation by cooling the product stream with water after compression to about 12 atmospheres. A flow diagram for this procedure is given in Figure 54.

Figure 54 shows that the C_5-C_8 fraction produced as a byproduct in the process is recovered by stabilising columns 6 and 7 and fed to the absorber 4 where it extracts the bulk of the C_4 product fraction. The unrecovered fraction at this stage is passed to column 7 as a reflux portion for stabilisation of the C_5-C_8 fraction and further recovery of the C_4 fraction. Similarly

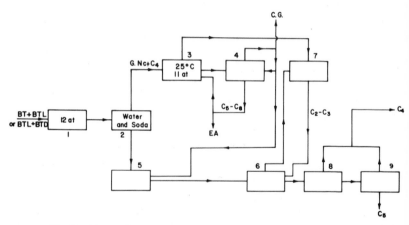

Fig. 54   Process for separating the butane-butylene-butadiene fractions.
1—compressor; 2—condensers; 3—absorption column; 4—desorption column; 5—reservoirs; 6 and 7—stabilizing columns; 8 and 9—distillation columns; EA stands for excess of absorbent; G for combustible gas and NoC for non-condensible gas.

in the desorber 4 a portion of the desorbed gases are refluxed to aid in desorption and further recovery of the heavier fractions. A portion of the distillate or desorbed gas stream is fed also to the condensate reservoir 5 and then along with the condensate to stabilising columns 6 and 7. The bottom fraction from column 7 is fed (as reflux) to the top of column 6 while the residue from column 6 is rectified in columns 8 and 9 to produce the heavy fraction (C_5 and above) and the C_4 or butadiene fraction.

The separation of *n*-butane from *n*-butylene in the C fraction is effected by extractive distillation because of the close relative volatilities of these gases.[368] The Bunsen coefficiénts for a number of good selective solvents for the separation of butylene from butane are sulpholone— 2·6, dimethyl sulphoxide—2·4, propylene carbonate—2·3, pyrrolidine— 2·3, nitrolactone—2·3, acetonitrile—2·3, dimethyl formamide—2·3, monothyl formamide—2·2, furfural—2·1. Acetone-water mixtures and N-methyl pyrrolidine also are good extractive solvents for this separation.

In practice the first step involves the separation of isobutane from isobutylene by extractive distillation while the isobutylene is separated from the mixtures of isobutylene and 1-butylene as well as isobutane and isobutylene mixtures by absorption in sulphuric acid. Extractive distillation with furfurol is normally employed to separate *n*-butane from 2-butylene.

The use of acetonitrile as an extractive solvent for the separation of C_4 has been reported to effect an increase of productivity of about 60%.[369] Presumably the same improvement would result also for other solvents whose separation coefficient was 2·3 or greater.

Although N-methyl pyrrolidine has a separation factor (2.1) which is somewhat lower than those for the solvents listed above, it has gained wide and large scale usage in recent years because it can be more economically utilised than cuproammonia acetate or other solvents for product streams containing high concentrations of butadiene. This is due primarily to the fact that with more intense desorption conditions a larger percentage of the acetylene and alene products react to produce polymers which impair the regeneration of the chemi-sorption solvents.

The selectivity of N-methyl pyrrolidine is adequate to permit a clean separation of butadiene from the C_4 fraction. The solubility of butadiene in N-methyl pyrrolidine at 20 and 40°C respectively is 87·8 and 43·0 m^3/m^3 atmosphere. The addition of 5-10% water to the pyrrolidine lower these solubilities by about 30%. The purity of butadiene obtained for degrees of recovery between 97-98% ranges from 99·5-99·9%. This corresponds to a butylene impurity of 0·1 to 0·5% in the butadiene.

The N-methyl pyrrolidine process also can be used to separate fractions of butane-butylene and butylene-butadiene from the dehydrogenation products of butane. It is also effective for separation of butane from butylene and butylene from butadiene.

The main disadvantages in employing N-methyl pyrrolidine in these separations are that energy (steam and electricity), solvent and gas recycle requirements are somewhat higher than with other solvents in particular where high product purity is the prime objective.[370]

The industrial separation of butadiene by chemisorption with cuprammonium acetate solutions is well described in the literature.[328,368,371–378] In principle the method consists of a reaction between butadiene and cuprous chloride in the presence of ammonium chloride with the formation of a yellow crystalline complex. Industrially, however, an ammoniacal solution of cuprous acetate is used. These are obtained by dissolving copper in ammoniacal solutions of ammonium acetate. A typical solution has the following composition 3·3 moles of copper (3 moles monovalent and 0·3 bivalent) 4 moles acetic acid, 11 moles ammonia and 3·3 moles of water per litre of solution.

The solubilities in moles per litre of butadiene, 1-butylene, 2-cis-butylene and 2-trans-butylene in the above solutions at 0°C and 0·5 atmospheres are 0·7, 0·07, 0·028 and 0·013 respectively. The respective relative volatilities at 4·55 atmospheres and 41°C on the other hand are 1·000, 1·040, 0·776 and 0·843.

If the chemisorption process is carried out at atmospheric pressure the selectivity of the solvent is significantly higher and installation costs can be reduced by about 50%. Electrical energy requirements also are reduced substantially since cooling with water only suffices at this pressure.

The cuprammonium acetate solution, however, as stated before is restricted to product streams with low (12-15%) content of butadiene.

The furfurol extractive distillation process is applied to the separation of butadiene from the butylene-butadiene fraction.[379] The furfurol to hydrocarbon ratios required to effect clean separation vary with the butadiene content of the product stream. At about 25% butadiene a ratio of 12:1 is used while at 50% butadiene in the hydrocarbon mixture the ratio is decreased to 6:1.

If the extractive distillation column contains about 150 trays a butadiene purity of 99·7% can be effected. The main impurity remaining in this case would be 2-trans-butylene.

Dimethyl sulphoxide also has been used as a solvent for industrial separation of butadiene because of its high selectivity for this hydrocarbon.[380] For example a product stream containing 42% butadiene, significant amounts of n-butane, isobutane, butylene and isobutylene, as well as 0·1% C$_5$ hydrocarbons, 0·4% vinyl acetylene and 0·1% ethyl acetylene, the top fraction from the dimethyl sulphoxide extractive distillation column contained a maximum of 0·3 % butadiene and the bottom fraction a minimum of 99·5 % butadiene. The chief impurity was *cis*-2-butylene and the combined impurity level of vinyl and ethyl acetylene did not exceed 100 ppm. The steam consumption was 2·7 tons per ton of butadiene.

In this case the paraffins and C_4 olefins are separated by extractive distillation with dimethyl sulphoxide in the first step and compounds with higher boiling points such as ethyl acetylene, 1,2-butadiene and C_5 hydrocarbons are separated by distillation in phase three.

The use of dimethyl formamide as a solvent to separate butadiene from C_4 fractions is applied in one of two ways.[381] The Leuna-Difex process of the VEB Machinen Und Apparatenbau firm employs dimethyl-formamide as a selective solvent in a single step. The Mitsubishi process of the Japanese Geon Co. Ltd on the other hand performs the extractive distillation in two steps. In the first, components less soluble than butadiene are removed and in the second the more soluble components are separated. A final distillation is required to separate butadiene from impurities with higher boiling points.

Both processes produce butadiene with 50 ppm or less of acetylene.

The purification of butadiene from C_4-acetylenes is of commercial importance, also. About 50-70 % of this fraction is vinyl acetylene which is produced in concentration of the order of 1000-5000 ppm when butadiene is manufactured by pyrolysis and 1500-2500 ppm when the butadiene is produced by dehydrogenation of butylene.

The acetylenic derivatives are eliminated in practice by liquid phase hydrogenation at 10-20°C and 2·5-7 atmospheres pressure using a noble metal catalyst on an inert support using a load of 20 kg C_4 per litre of catalyst. This catalyst can be regenerated with oxygen at 400-500°C.[382]

A modification to the above process consists of gradually adding the hydrogen to at all times avoid saturation with hydrogen. In this way due to the higher reactivity of vinyl acetylene the losses of butadiene are held down to 1-2 % or less. The selectivity is greater for vinyl acetylene because the first hydrogenation product is 1-3-butadiene.

The above procedures are applied prior to the extractive distillation procedures discussed earlier. After extractive distillation the acetylenic derivatives may be eliminated by absorption solutions of ethanolamine containing cuprous chloride. Here butadiene losses are only 0·7–0·8 %.

DEHYDROGENATION OF ISOPENTANE AND ISOPENTENE

Before discussing the industrial dehydrogenation process for the production of isoprene,[383] it is of interest to look at other processes which also are used on an industrial scale and which appear to be competitive in some aspects with it.

The Italian SNAM process is among these.[384] In this process the isoprene is produced by condensation of acetone with acetylene (i.e. the ethynylation of acetone) with the formation of methyl butynol, its conversion to methyl butenol and dehydration of the latter to isoprene.[385-388] The total yield is about 89 % and the isoprene obtained is of high purity however the raw material costs for this process are significantly higher than those for other processes.

In the French process isoprene is obtained by condensing isobutylene with formaldehyde and catalytically cracking the resultant product dimethyl-meta-dioxan.[389-391] The catalytic cracking is carried out in a fluidised bed reactor with a yield of about 77 % based on isobutylene and about 48 % based on methanol required for the formation of formaldehyde. Formaldehyde and water are the major co-products of this process hence the purity of isoprene attainable is high.

The American (Goodyear) process produces isoprene by dimerising propylene to 2-methyl pent-1-ene, isomerising the latter with a yield of 99 % to 2-methyl pent-2-ene and converting this isomer by pyrrolysis to isoprene and methane.[392-394] The yield based on propylene is about 50 %. This process could be highly competitive if adequate direct isomerisation

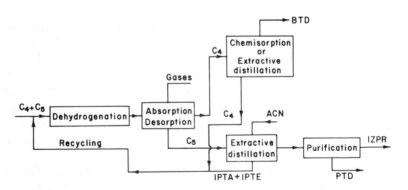

Fig. 55   Process for simultaneous dehydrogenation of C_4 and C_5 fractions.
ACN—aqueous acetonitrile solution; IPTA—isopentane; IPTE—isopentene; IZPR—isoprene; PTD—1-3-isopentadiene (pyperilene); BTD—butadiene.

were achieved. The dehydrogenation of the C_5 fraction however still holds a raw material cost advantage.

Isoprene also has been produced on a pilot plant scale by Houdry with a 50% yield by reacting a mixture of isopentane and isopentene over a chromium oxide catalyst on an alumina support at 600°C and subatmospheric pressures.

The dehydrogenation process employed by Shell at two industrial installations is similar to that for the production of butadiene. The raw material in this case isopentene is reacted with superheated steam at 700°C over a catalyst comprised of calcium phosphate and nickel to produce isoprene.

A new alternative to the Shell process called the IDAS process employs iodine to perform an oxidative dehydrogenation of the isopentene. The hydrogen iodide formed is recovered via reaction with nickel oxide to form nickel iodide and the iodine is regenerated from the latter by oxidation with air at 500°C. Higher conversions and yields are obtained by this process at lower temperatures, in particular with isopentene as the feed material. However, this would hold true also if the feed were comprised of a mixture of isopentane and isopentene (see the process tested by Houdry above) since successive oxidative dehydrogenations of paraffin to olefins to diolefins are more efficient than dehydrogenations of paraffins to diolefins carried out in a single step.

The simultaneous dehydrogenation of mixtures of isopentenes and butylenes is of special practical importance.[395-398] In this process butadiene and isoprene are produced simultaneously in the reactor in a ratio determined by the composition of the feed stream. The dehydrogenations are effected at about 600°C and 250-400 mm Hg pressure over a catalyst comprised of iron oxide, chromium oxide and potassium carbonate. After dehydrogenation the C_4 and C_5 fractions are separated by absorption-desorption and the paraffinic and olefinic C_4 and C_5 fractions are separated by extractive distillation and recycled to the reactor. The butadiene product obtained thus is of adequate purity while the isoprene contains significant levels of piperylene and requires a further purification step. The essential process details as described are depicted in Figure 55. Generally the fresh feed is comprised of about 87% butane, 16% isopentane, 3% isobutane and n-pentane while the recycle stream will contain about 47% n-butane, 26% n-butylene, 12% isopentane, 7% isobutylene, 7% isobutane and isopentene and 0·1% isoprene, butadiene, etc.

CHAPTER 5

Ammoxidation of Olefins

5.1 ACRYLONITRILE

5.1.1 General Overview

At the present time acrylonitrile is one of the most important acrylic monomers due primarily to the very fast development of a variety of acrylic fibres. Its major uses are in the manufacture of plastomers, elastomers and certain organic compounds. Rhone-Poulence, Imperial Chemical Industries, Union Chemique Belge and Monsanto are continuing intensive research on the production of adiponitrile, the basic material for the manufacture of 6,6 nylon, starting from acrylonitrile.

In 1970 in Europe, 77% of the acrylonitrile consumed was used in the manufacture of fibres, 15% for plastics materials, 7% for synthetic rubber and 1% for miscellaneous products. The pattern of usage suggests that the proportion allocated to manufacture of fibres and rubber will decrease in future in favour of increasing usage in the manufacture of plastics.

The miscellaneous uses although small in tonnage are nonetheless of major importance. For example the use of acrylonitrile to modify the properties of compounds containing an active hydrogen as characterised by amines and alcohols holds special significance in the synthetic fibre industry. The β-cyanoacetylation of cellulose for example causes moderate swelling of the fibre which imparts greater thermal and chemical stability to it and makes it easier to dye. Similarly cyanoacetylation of aromatic amines produces intermediates which have wide utility in the production of dispersion dyes for polyester and other synthetic fibres.

5.1.2 Manufacturing Processes

The processes used for the manufacture of acrylonitrile on a worldwide basis may be divided into two main groups: that based on ammoxidation of propylene and those based on another principle and/or on another raw material.

140

In the former process acrylonitrile is produced by reacting propylene, ammonia and air over an appropriate catalyst. The overall reaction can be summarised as:[407-412]

$$CH_2{=}CHCH_3 + NH_3 + \tfrac{3}{2}O_2 \;\rightarrow\; CH_2{=}CHCN + 3H_2O$$

The details about this reaction and about the technological considerations such as type of catalyst, reactor and performance are given in Sections 5.1.3 and 5.1.4 respectively.

Other processes which also will be discussed more fully later are as follows.

The vapour or liquid phase catalysed reaction between acetylene and hydrogen cyanide according to the reaction, $CH{\equiv}CH + HCN \rightarrow CH_2{=}CHCN$, was first developed in 1940 by I. G. Farben in Germany. In 1952 Monsanto put this process on line in the USA, and this was followed by American Cyanamid in 1953 and Goodrich in 1955.

In the liquid phase, a molar ratio of 10:1 for acetylene:hydrogen cyanide is used and the reaction is carried out at about 70°C at atmospheric pressure in the aqueous catalytic solution which is comprised of 26% cupric chloride. About 20 kilograms of acrylonitrile are produced per kilogram of dissolved copper before the catalyst needs to be regenerated. The regeneration of the catalyst is generally effected by precipitating the copper with zinc and then reconverting it to cupric chloride. The reported yields based on acetylene are about 80% and on hydrogen cyanide about 85-90%. The product purity is high being about 99%.

In the vapour phase this reaction has been carried out at 400-500°C over a catalyst comprised of metal cyanides in fixed bed reactors using equal volumes of acetylene and hydrogen cyanide feed. At present this approach does not constitute an industrially competitive process. According to some researchers the du Pont process offers improvements over the previous process in that a more active catalyst is employed resulting in improved product yield and quality particularly with respect to degree of contamination by secondary products.

Since 1960 anhydrous catalyst systems based on cupric chloride also have been studied with considerable interest and success. Adiponitrile and pyrrolidine are effective substitute solvents to water and in this event the reaction conditions are about equal to those used with the aqueous catalyst.[413]

The Knapsack process based on the preparation of lactonitrile from acetaldehyde and hydrogen cyanide and its subsequent dehydration, also, is of interest here. The pertinent reactions of this process are;

$$CH_3CHO + HCN \xrightarrow[\text{pH 7-7·5}]{\text{NaOH} \quad 10\text{-}20°C} CH_3CHOHCN \qquad \text{(Ref. 413)}$$

$$CH_3CHOHCN \xrightarrow[\text{600-700°C}]{\text{H}_3\text{PO}_4} CH_2{=}CHCN$$

or

$$CH_3CHOHCN \xrightarrow{(CH_3CO)_2O} CH_3CH(CH_3COO)CN$$

$$CH_3CH(CH_3COO)CN \xrightarrow{540\text{-}580°C} CH_2{=}CHCN + CH_3COOH$$

$$\text{(Ref. 414)}$$

This process was introduced commercially in 1958-59 and was considered to be the first effective improvement to the industrial processes existing at the time resulting in substantial reduction in the manufacturing cost of acrylonitrile. Bayer in 1959 and BASF in 1962, abandoned their initial processes based on the acetylene-hydrogen cyanide reaction and replaced these with Knapsack technology.

The main advantages of the Knapsack process are relatively high yields, good purity of crude product and high purity of finished product. The possibility of obtaining acetaldehyde directly from ethylene economically (as outlined earlier in the chapter on liquid phase oxidation reactions) further improved the opportunity to extend the use of this process. The acrylonitrile produced by this process contains about 0·005 % divinyl acetylene and less than 0·12 % methyl vinyl ketone as the major impurities.

Another recent process which can be considered as a variation of the ammoxidation process is that developed by the du Pont Co. which employs nitrous oxides (*in lieu* of ammonia and air) to react with propylene to form the acrylonitrile.[415] The overall reaction in this case is

$$4CH_2{=}CHCH_3 + 6NO \rightarrow 4CH_2{=}CHCN + N_2 + 6H_2O$$

The nitrous oxide reactant gas contains about 15 % NO and is produced by catalytic oxidation of ammonia. The propylene to nitrous oxide ratio in the feed is about 5 to 1. Although the conversion per pass is low, adequate productivity is achieved at normal pressures in multi-tubular reactors. The fixed bed catalyst is one based on silver deposited on silica gel which is used at 460-500°C without a promoter and at 420-500°C with calcium oxide as the promoter additive. The heat of the reaction is recovered by circulating molten salts through the reactor and heat exchangers.

From the available data as summarised in part in Table 21, it is clear that on a world basis only about one-seventh of the acrylonitrile is manufactured by processes other than that based on the ammoxidation of propylene. The following production quotes are comparable and indicative of the reason for this preference. For an installation with an annual capacity of 18 130 tons, the ammoxidation of propylene process has a base cost of $340·00 per ton of acrylonitrile, whereas the cost per ton of acrylonitrile produced from acetylene and hydrogen cyanide for an installation with an annual capacity range from 18 120 to 45 300 tons ranges between about $840·00 to $1200·00. The latter figures include the costs for acetylene production, also.

TABLE 21
World production of acrylonitrile (in thousands of tons[416])

Country	Production (beginning of 1970)	Planned expansions
West Germany	110	+170 at end of 1972
France	90	+60 at end of 1972
Holland	45	+45 at end of 1971
Italy	92	+60 at end of 1972
England	40	+180 at end of 1972-73
USA	651	+196 in 1970
Japan	311	+316 in 1971
Bulgaria	22	—
East Germany	35	—
USSR	75	—

In 1970 the Monsanto Co. closed its 59 000 tons per year acrylonitrile plant in Texas City based on the acetylene hydrogen cyanide process and it seems most probable that the Montedison Co. also will close its 30 000 tons per year plant for acrylonitrile production by this process. Other similar closings may be anticipated in future.

It is evident from Table 21 that the worldwide production of acrylonitrile as planned will almost double between 1971 from 1 530 000 tons per year to 2 750 000 tons per year by 1973. This increase will be established primarily by the ammoxidation of propylene. Current world statistics and forecasts on the production of acrylonitrile show that the almost doubling in production has indeed occurred in the past several years and barring setbacks due to shortages of propylene arising from the present oil crisis a growth rate of 10 to 15% per year is likely to continue for the next several years.

Although the Knapsack process could be competitive with the process based on the ammoxidation of propylene it does not have the potential at present because the availability of propylene from existing (and planned installations surpasses by a considerable margin that of ethylene.

5.1.3 Kinetic and Technological Considerations

MAJOR CHEMICAL REACTIONS OF THE AMMOXIDATION PROCESS

Although other reactions may and do occur when propylene is reacted with ammonia and air over catalytic materials, the major primary and secondary reactions may be summarised as follows:

(a) $C_3H_6 + NH_3 + \frac{3}{2}O_2 \xrightarrow[-3H_2O]{} CH_2=CHCN$

$$\Delta H = -123 \text{ kcal/mol}$$

(b) $\frac{2}{3}C_3H_6 + NH_3 + O_2 \xrightarrow[-2H_2O]{} CH_3CN$

$$\Delta H = -86 \cdot 6 \text{ kcal/mol}$$

(c) $\frac{1}{3}C_3H_6 + NH_3 + O_2 \xrightarrow[-2H_2O]{} HCN \qquad \Delta H = -75 \text{ kcal/mol}$

(d) $\frac{1}{3}C_3H_6 + \frac{3}{2}O_2 \longrightarrow CO_2 + H_2O \qquad \Delta H = -153 \text{ kcal/mol}$

(e) $\frac{1}{3}C_3H_6 + O_2 \rightarrow CO + H_2O \qquad \Delta H = -85 \cdot 5 \text{ kcal/mol}$

(f) $C_3H_6 + O_2 \xrightarrow[-H_2O]{} CH_2=CHCHO \qquad \Delta H = -84 \cdot 4 \text{ kcal/mol}$

(g) $\frac{2}{3}C_3H_6 + \frac{1}{2}O_2 \rightarrow CH_3CHO \qquad \Delta H = -43 \text{ kcal/mol}$

(h) $C_3H_6 + NH_3 + O_2 \xrightarrow[-H_2O]{} CH_3CH_2CN$

$$\Delta H = -98 \cdot 4 \text{ kcal/mol}$$

(i) $C_3H_6 + \frac{1}{2}O_2 \rightarrow CH_3COCH_3 \qquad \Delta H = -56 \cdot 7 \text{ kcal/mol}$

(j) $C_3H_6 + \frac{3}{2}O_2 \xrightarrow[-H_2O]{} CH_2=CHCOOH$

$$\Delta H = -146 \cdot 5 \text{ kcal/mol}$$

(k) $\frac{2}{3}C_3H_6 + O_2 \rightarrow CH_3COOH \qquad \Delta H = -109 \text{ kcal/mol}$

(l) $CH_2=CHCHO + NH_3 + \frac{1}{2}O_2 \xrightarrow[-2H_2O]{} CH_2=CHCN$

$$\Delta H = -38 \cdot 6 \text{ kcal/mol}$$

THE TWO STEP ASPECT OF THE PROCESS

Initially when this process was being developed the process was carried out in two steps. In the first step acrolein was produced in accord with reaction (f) above and this intermediate then was reacted further with ammonia and air to produce acrylonitrile in accord with reaction (l) above.

It is probable that the second step forms an addition compound of the aldehyde-ammonia type as an intermediate product which by the double process of oxidation-dehydration (at the high reaction temperatures) is converted to acrylonitrile. Although it is expected that the formation of the addition compound would be fast and faster than the subsequent oxidation-dehydrogenation reaction, this intermediate has not been isolated to date. However the presence of small amounts of trimers, tetramers, cyclic trimers and other low molecular weight polymers are believed to provide proof of the formation of the intermediary addition compound of the aldehyde-ammonia type.

The following observations are pertinent to the two step process. The rate of formation of acrylonitrile depends upon the partial pressure of acrolein if ammonia is present in excess and on the partial pressure of ammonia if acrolein is in excess. Both of the foregoing observations apply only when oxygen in excess of the requirements for the reactions is present. At low and constant partial pressures of both ammonia and acrolein the reaction is independent of the oxygen concentration when it is present with an excess of about 10% and holds a first order dependence on oxygen at low concentration of the latter.

The transformation of acrolein to acrylonitrile (reaction (1) above) at 380°C over a catalyst comprised of about 25% molybdenic acid on an alumina support occurs with the following yields as a function of contact time.

Contact time (seconds)	Yield of acrylonitrile (%)
0·05	30
0·10	50
0·15	60
0·20	65
0·25	70
0·30	72
0·35	75
0·40	77
0·50	79
0·60	80
0·70	80

ASPECTS OF THE ONE STEP PROCESS

The following specifications apply to the modern processes in which acrylonitrile is produced by direct reaction of a mixture of propylene ammonia and air in a single step.

The rate of the reaction (r) depends directly on the propylene concentration (c) in accordance with the expression $r = kc$, where k is the reaction rate constant, at partial pressures of oxygen in excess of 0.03 atmospheres.[417,418]

Determinations of the yield of acrylonitrile as a function of contact time also have been made for this process. For a reaction mixture comprised with the ratio of $C_3H_6 : O_2 : NH_3 : H_2O(v)$ of $1 : 2.7 : 1.1 : 1$ at a total pressure of 800 mm Hg and a reaction temperature of 465-475°C it was found that the reaction rate constant calculated using the first order dependence and the relationship between conversion and contact time varied between 0.286 and 0.298 seconds^{-1} for contact times of 1 to 5 seconds. Thus it was concluded that the transformation of propylene to acrylonitrile proceeds by an apparent first order reaction. On the basis of this conclusion and the reaction rate determinations at various temperatures the classic Arrhenius

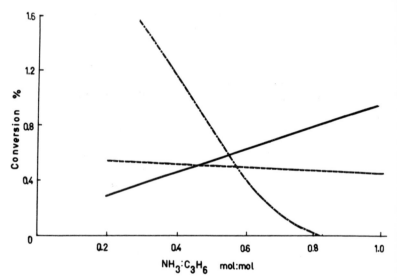

Fig. 56   Influence of molar ratio of $NH_3 : C_3H_6$ on the ammoxidation reaction for contact time of 2 seconds at 450°C using a reaction mixture ratio of $C_3H_6 : O_2$ (from air): H_2O of $1 : 1.5 : 1$.
——— conversion to acrylonitrile; —·—·— conversion to acrolein (with ordinate value times ten); – – – – total conversion of propylene.

activation energy for the overall reaction was calculated to be 18 300 calories.

Higher $NH_3:C_3H_6$ ratios (and consequently lower C_3H_6 concentrations) were found to have only a slight influence on the total conversion of propylene as depicted in Figure 56. The conversion to acrylonitrile on the other hand increases quite sharply as the $NH_3:C_3H_6$ ratio is increased. Figure 56 shows that at $NH_3:C_3H_6$ ratios above about 0·8 acrolein does not appear, which is important to industrial operations. The presence of acrolein leads to the formation of cyanohydrin and cyanohydric acid which complicate the separation and purification steps.

The total conversion of propylene also increases with an increasing molar ratio of $O_2:C_3H_6$ but within narrow limits as shown in Figure 57. The selectivity of propylene transformation to acrylonitrile relative to total propylene reacted also increases sharply within similar narrow limits. Moreover the conversion to acrylonitrile is more accentuated at low space

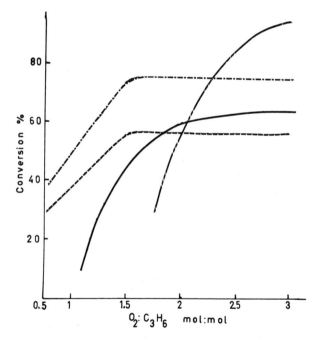

Fig. 57   Influence of molar ratio of $O_2:C_3H_6$ on the ammoxidation reaction.
—— total conversion of propylene; — — — — — selectivity; — · — · — conversion to acrylonitrile at 11 l/l.cat.h; — — — — conversion to acrylonitrile at 40 l/l.cat.h.

velocities. Thus although these are not shown in Figure 57 the conversions to acrylonitrile at space velocities of 20, 60, 100 and 140 litres per litre of catalyst per hour were found to be 72, 55, 38 and 28 % respectively for the conditions as specified.

At a molar ratio of $O_2:C_3H_6$ of about 0·016, deactivation of the catalyst occurs presumable due to the oxidation reactions removing oxygen from the lattice.

The rate of the reaction as to be expected increases with the temperature, however the total conversions and yields of the various products generally pass through maxima as the temperature is raised. Four typical conversion curves are shown in Figure 58. In this figure the reaction mixture mol

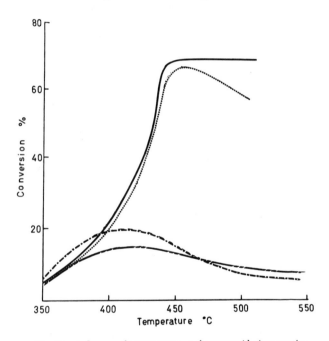

Fig. 58   Influence of temperature on the ammoxidation reaction.
—— total conversion of propylene; – – – – conversion to acrylonitrile; —·—·— conversion to acetonitrile; – – – – – – conversion to hydrogen cyanide.

ratio for $C_3H_6:NH_3:O_2:H_2O$ was 1:1:1·8:1. At 400, 440 and 500°C respectively the total conversion of propylene can be read off as 20, 68 and 70 %. Above certain temperature limits the conversion to acrylonitrile increases while that to acetonitrile and cyanohydric acid decreases.

However at still higher temperatures the conversion to acrylonitrile may decrease again. Thus at 450°C a conversion of 68 % is shown in Figure 58, whereas at 500°C this value falls to about 60%.

CATALYSTS

Among the various catalysts used in the ammoxidation of propylene the first that performed well were those based on oxides of bismuth and molybdenum. Initially the oxide system employed was of the type $Bi_2O_3.(MoO_3)_n$ which according to Russian scientists[418,420] is comprised of three compounds with distinct phases; $\alpha \sim Bi_2O_3.3MoO_3$, β-Bi_2O_3. $2MoO_3$ and γ-Bi_2O_3-MoO_3. Researchers in Holland[419] however disagree with the Russian findings with respect to the existence of the β-form. They report instead the existence of the compound $3Bi_2O_3.MoO_3$.

The initial studies have shown that the β form (or compounds comprised of the oxides with a Bi:Mo ratio of 1:1) has maximum catalytic activity. This may be accounted for by the fact that MoO_3 alone reduces the selectivity for acrylonitrile formation while Bi_2O_3 alone does not catalyse the propylene ammoxidation reaction. The catalyst activity or rather performance as a function of the Bi:Mo atomic ratio is tabulated in Table 22.

TABLE 22
Performance of $Bi_2O_3.MoO_3$ catalyst system as a function of atomic ratio of Bi:Mo

Atomic ratio Bi:Mo	Specific surface (m^2/g)	Total conversion (%)		Percent conversion C_3H_6 reacted to				
		C_3H_6	NH_3	ACRN	ACTN	ACR	HCN	CO_2
0·1	2·0	15·7	64	21	15	15	10·5	7
0·07	2·8	43·5	73·5	60·5	4·2	3	3·5	5·5
0·46	2·5	55	85	71	3·6	—	2·4	4·7
0·67	2·8	62·5	81·5	71·4	2·5	—	1·7	5·0
0·92	2·5	66·4	84·3	71·4	2·0	0·7	2	4·8
1·15	3·0	68·7	80·3	72	1·5	0·8	0·8	5·6
4·60	2·6	17·0	53·3	4·3	1·3	—	0·5	−89
2·0	2·0	8·8	17·5	0	—	—	0·6	89

It is evident that increasing the Bi:Mo atomic ratio to a value of 2 or greater causes the catalyst to function as a total oxidation catalyst. Also the maximum activity and selectivity for acrylonitrile formation, as stated earlier, can be seen from Table 22, to be found in catalysts with a Bi:Mo atomic ratio of about 1.

Studies on catalyst improvement led to the development of the bismuth phosphomolybdate type.[421,422] The performance of these was appreciably greater than that of the straight oxide type but their mechanical resistance was low. A method of preparation of this type of catalyst has been reported which minimises the internal strain of the particles, thereby improving their mechanical resistance.[423]

The procedure employed to impart greater mechanical strength to the bismuth phosphomolybdate type catalyst was as follows. Samples of this catalyst with a particle size of about 7 microns were mixed with particles of corundum with a particle size of about 12 microns, wet ground and pelletised. Agglomeration and dense packing of the pellets was achieved by employing vibrating devices with frequencies of 10 000, 15 000 and 20 000 cycles per minute and amplitudes of 0·05 to 0·3 mm. Optimum results were obtained at 10 000 cycles per minute and an agglomeration time of 4 to 5 seconds. These agglomerated samples were then compacted in a hydraulic press and then preheated at $550 \pm 10°C$ (the usual working temperature of the catalyst) for 6 hours prior to evaluation.

It was found that samples which had been agglomerated (or precompacted) through vibration could be compressed to the same bulk density as samples prepared by static compaction using press pressures 100 times smaller. Moreover under static compression durable pellets with bulk densities in excess of $3·75 \text{ g/cm}^3$ could not be made whereas with vibro-agglomeration, pellets with high mechanical strength and bulk densities up to $4·0 \text{ g/cm}^3$ were readily obtained. Indeed it was concluded that the vibro-agglomeration technique permits the particles to settle optimally thereby effecting a close approach to maximum packing with minimum internal strain. The mechanical resistance of statically agglomerated samples was 220 kg/cm^2 at best whereas that for vibro-agglomerated samples was 450 kg/cm^2.

The conversions and yield of acrylonitrile from a reactant mixture comprised of 10% (by volume) propylene, 10% ammonia, 15% oxygen, 10% water vapour and 55% nitrogen were determined for these catalysts at 480°C and a contact time of 2 seconds. No differences in conversion or yield were found for the catalysts tested irrespective of their method of compaction and mechanical strength. However after 500 hours of operation the mechanical resistance of the high density $(3·9-4·0 \text{ g/cm}^3)$ vibro-agglomerated catalyst pellets was reduced by only about 25% whereas that of the statically agglomerated pellets was reduced to zero.

Tests with additives such as K_2O, Li_2O, BaO, Al_2O_3, Fe_2O_3, CuO, NiO and Cr_2O_3 have shown that these oxides provide little to no promoter activity. Indeed additives based on Li, K, Mn, Ba, Zn, and Ag reduce the catalyst activity whereas additives based on Cu, Pb and As tend to promote the total oxidation reaction.

The optimum support for the bismuth phosphomolybdate catalyst appears to be silica gel. Other supports which have been evaluated include alumina, bentonite, kieselguhr, and metal oxides with good thermal conductivity. The activity of this catalyst on a silica gel support with respect to conversion of propylene to acrylonitrile has been found to be 100 times greater than when bentonite is used as the support, 14 times greater than with an alumina support and 90 times greater than with a kieselguhr support.

In summary then, the most effective catalyst developed to date based on Bi and Mo is the bismuth phosphomolybdate on a silica gel support.

Other catalyst systems which have been studied in recent years include those based on oxides of antimony and iron and others based on zeolites. Catalysts prepared by precipating the hydroxides of antimony and iron from acid solutions of antimony chloride and iron nitrate with ammonia have been shown to be comprised of the individual oxides Sb_2O_4 and Fe_2O_3 as well as the compound $FeSbO_4$.[424]

The activity and selectivity of the catalyst system based on oxides of iron and antimony were determined under isothermal conditions in fixed and fluidised bed reactors using a reactant mixture comprised of 7% (by volume) of propylene, 8·5% ammonia, 16·5% ammonia and 68% nitrogen. The pertinent conclusions of these studies are the following.

Iron oxide is the more active component of this catalyst system, however, in its presence alone, only about 2·5% acrylonitrile but 93% carbon dioxide is formed. The activity of antimony oxide, on the other hand, is low but its selectivity is high; about 41% of the total reaction products formed is acrylonitrile. In the composition range 56-97% Sb_2O_4, the selectivity for acrylonitrile formation reaches the maximum of about 80%. Since the compound $FeSbO_4$ has a high selectivity for acrylonitrile formation it is natural to conclude that the selectivity of the $Fe_2O_3.Sb_2O_4$ catalyst is determined largely by its $FeSbO_4$ content. It should be noted here that antimony is used also with success in other catalyst systems for the ammoxidation of propylene.

In the Sohio-USA process the catalysts employed are based on the oxide, molybdate or phosphomolybdate of antimony. One of the types of catalysts employed in the Distillers-Anglia process is antimony oxide. The

Knapsack process also uses catalysts based on phosphates and oxides of antimony.

Recently researchers in the USSR have studied zeolites with various degrees of ionic substitution by Na^+, Ca^{2+} and Fe^{3+} ions. Three types classed as type A, X and Y were used.[425] They found that the intensity of the oxidation reaction depends on the position of the iron ions in the crystalline lattice. The type of zeolite used was found to be of particular significance. Thus for the different structural types of zeolites evaluated for catalytic activity with respect to the ammoxidation process for the same degrees of substitution (30-55 %) of the sodium ions by iron ions the activity of type X was found to be greatest, type Y in between and type A, least.

The replacement of three sodium ions by one ferric ion increases the spatial volume for absorption and this may account for the simultaneous increase in catalytic activity. Altering the degree of substitution of the sodium and/or potassium ions in the zeolites with ferric ions also alters the yield of acrylonitrile obtained with the thus modified zeolite catalysts.

The highest activities and selectivities were found for zeolites of the FeNaX, FeCaX and FeCaA type in which about 50 % of the sodium or calcium ions had been replaced by ferric ions. The productivity of these iron-zeolite catalysts, however, was not at all comparable to that of the bismuth-molybdenum based catalysts. As an example at 480°C the selectivities and productivities for the FeNaX zeolite catalyst and the bismuth-molybdenum based catalyst were 26 % and 30 grams of acrylonitrile per litre of catalyst per hour relative to 75 % and 180 grams of acrylonitrile per litre of catalyst per hour, respectively.

Zeolites of type Y and X with about 50 % of their ions replaced by ferric ions were found to favour total oxidation in preference to the ammoxidation reaction. Thus although the performance of zeolite type catalysts can be improved somewhat by replacing some of their lower valent ions with ferric ions, it is apparent that without more substantial improvements (through other modifications) they cannot compete with the catalysts currently utilised in manufacture of acrylonitrile by the ammoxidation of propylene.

Finally brief mention must be made of the catalysts based on uranium which are most widely used in the modern Sohio-USA process which appear to yield the optimum performance. A recent study[426] of the chemistry of interaction between antimony oxides and uranium oxides has shown that two crystalline phases of special importance to the catalytic process are formed. The first phase which initially was considered to be

$(UO_2)Sb_3O_7$ was proven to be USb_3O_{10} and the second phase $USbO_5$ rather than $Sb_3U_3O_{14}$.

The crystalline structure of phase one (which was obtained from a reaction mass composed initially of antimony oxide and uranium, by dissolving the excess antimony oxide in hydrochloric acid) was determined by comparing its X-ray diffraction pattern with that assumed for USb_3O_{10} and other related systems. The phase two compound $USbO_5$ is structurally similar to phase one. This can be seen from a comparison of the respective diffraction patterns and by the fact that by replacing antimony with uranium at the equivalent point of the spatial group of USb_3O_{10} the compound $U_2Sb_2O_{10}$ or $USbO_5$ results. The positions of the atoms in this phase although close to the equivalent positions of phase one, are nonetheless somewhat displaced leading to a generally distorted crystalline sub-group of lower symmetry. The phase USb_3O_{10} has orthorhombic symmetry and in accordance with the general tendency of oxygenated compounds of uranium and antimony the coordination numbers for these two metal ions are 8 and 6, respectively.

5.1.4 Technological Aspects

THE TWO STEP AMMOXIDATION PROCESSES FOR PROPYLENE

The two step process developed by Distillers Co. in 1955 established the basis for the Distillers-Ungine single step process and was the first to prove out the possibilities and advantages of replacing ethylene and acetylene in the manufacture of acrylonitrile by propylene.[415]

In the first step of this process, as stated earlier, propylene is oxidised to acrolein. In the second step the acrolein reacts with ammonia and air at relatively high temperatures over a catalyst of molybdenum oxide on a silica or alumina support to form the acrylonitrile. With silica supports the $MoO_3:SiO_2$ ratio is generally about 20:80 and in the case when phosphoric acid is added (as a promoter) the catalyst contains phosphomolybdenum complexes.

The air to propylene ratio in the first step is generally 10:1 and the optimum temperature range lies between 250 and 350°C. The preferred catalyst is based on oxides of copper on a silica or alumina support. Small amounts of selenium also are generally added to the reaction mixture. The acrolein produced is passed directly to the ammoxidation reactor without preliminary separation or treatment.

For the second step of the process the ammonia to acrolein ratio is adjusted to 1·1-1·5:1 and the oxygen to acrolein ratio to 0·5-1:1. In addition to the nitrogen diluent which accompanies the oxygen in the air stream, steam also is added. If the concentrations of oxygen comprise 3 % or less of the reactants the explosion limits are not approached and the life of the catalyst is extended considerably. The acrolein, air and steam are mixed in the proportions specified earlier and preheated (if necessary) above 120°C before adding the ammonia. This is essential to prevent the formation of resinous products at this stage. In the case of a fixed bed multitubular reactor the feed stream temperature is generally about 350°C and the reactor temperature 380-400°C.

When the catalyst is fresh small amounts of hydrogen cyanide, aceto-nitrile and carbon dioxide are formed. As the catalyst ages the percentage of these secondary products formed decreased slightly.

THE SINGLE STEP AMMOXIDATION PROCESS

As to be expected the two step ammoxidation process has been largely replaced by the single step process which was first developed by the Standard Oil Co. of Ohio with the first industrial installation commencing operation in 1960. Somewhat later in 1965-1968, Montecatini-Edison SPA, Sohio-Badger Co. Inc. OSW, Distillers-Ungine, Bayer, SNAM and others followed suit by developing and putting similar processes into production.

The primary differences between the processes of these various firms lie in the characteristics of the catalyst, the type of reactor employed, the way in which the reactor temperature is controlled and the heat of the reaction utilised, the concentration of the reactants and method by which they are diluted as well as other minor but important technological operating parameters.

The chief characteristics of a number of single step ammoxidation processes are listed in Table 23.

Among the major problems which the ammoxidation reaction presents at the commercial level is the need to quickly and effectively remove the exothermic heat of the reaction and the related problem of chemical and mechanical stability of the catalyst. In fluidised bed reactors a high degree of temperature control and uniformity can be readily effected with a resultant improvement in the degree of conversion and yield. However in this type of reactor catalyst losses are high.

The fixed bed reactor, although better known and more reliable in operation, is nonetheless more difficult and complicated to operate satisfactorily. This is because of the need to design and operate the reactor as a molten salt heat exchanger with fine control over narrow temperature limits. In this case although the degree of conversion may be high yield losses of acrylonitrile can be substantially greater than in a fluidised bed reactor. To improve the heat exchange rate and capacity, the reactants are diluted with 60-80 % of inert materials (nitrogen and steam). This measure also is effective in reducing the possibilities of hot spots being developed in the catalyst tubes and as stated earlier in lowering the oxygen content to well below the explosion limits of the feed stream. These operating conditions favour long catalyst life but clearly reduce the productivity or production capacity of a given size installation. It should be noted from Table 23 at this stage that in the SNAM process 20 mols of steam are used per mol of propylene to effect the desired operating characteristics.

In the Montecatini-Edison SPA process the heat exchange elements are positioned within close proximity of the grid in the fluidising zone of the reactor. In the OSW process (Osterreichische Stickstoff-Werke A.J.) the fixed bed reactor is similar to that utilised in the manufacture of phthalic anhydride. A small scale reactor may be comprised of about 60 pipes, 2 metres long and 25 mm in internal diameter. In such a reactor the reactants would pass first through a layer of corundum granules about 60 cm deep, 3-4 mm in size and coated with a layer of the catalyst about 0·05 mm thick. They would then pass through a second layer of catalyst granules which in this case would be about 100 cm deep, 4-5 mm in size and coated with a layer of the catalyst about 0·3 mm thick.

For each kilogram of acrylonitrile produced at 460°C there are 3200 kilocalories of energy released, which must be removed.[433] This heat of the reaction is used to generate high pressure steam using in the latter case molten salts as the circulating heat exchange medium. The energy recovered in this way may be used to generate electrical power, distillation energy and/or steam diluent for the reaction. Thus of the 14·5 tons of steam required per ton of acrylonitrile in the OSW process, 6 tons are produced from the energy recovered from the reactor.

In the Distillers-Ungine process a fixed bed reactor is used, thereby ensuring low losses of catalyst, and by employing appropriate heat exchangers the heat of the reaction is used to generate process steam at 14 atmospheres. Also by employing a cooler on the reactor effluent stream additional process steam at 3 atmospheres is generated.

The catalysts generally contain 70-75 % support or inert material, with silica being the predominant material used. The catalyst life is determined both by the nature of the catalytic material and the type of reactor used. In a fluidised bed reactor catalyst life is generally shorter because of the

TABLE 23

Comparative data on various modifications of the single step propylene ammoxidation process

Process	Montecatini-Edison SPA	Sohio Badger Co. Inc.	OSW	Distillers-Ungine	Bayer	SNAM
References	427	414, 427	428, 432, 433	429, 430	415	415, 431, 434-438
Catalyst type	Te-Ce-Mo on SiO_2, in ratio of 20-30:80-90	U, Sb	Bi-Mo on corundum granules	Sb, Sn	B, P and traces of Fe, Bi, Mo, Al	V-Mo-Bi
Reactor type	fluidised bed	fluidised bed (high conversion per pass)	fixed bed	fixed bed	fluidised bed	fixed bed (multitubular with molten salt exchanger)
$C_3H_6:NH_3:Air:$ steam (mol:mol in feed)	1:1:11:-	—	1:1:8:1-2	—	1:1:7:1-5	1:1:1:8:20
Reaction temperature (°C)	420-460	400-510	450	370-480	480-540	500-520
Pressure (atmospheres)	~2	0.5-2	—	—	—	—
Purity of propylene (%)	92-93	—	>90	>90	—	—
$C_3H_6/ACRN$ (tons/ton)	1.15-1.30	0.99	1.25	1.35-1.45	—	—

	Col 1	Col 2	Col 3	Col 4	Col 5	Col 6
NH_3/ACRN (tons/ton)	0·55	—	0·5	0·55-0·60	—	—
HCN/ACRN (tons/ton)	0·060	0·08	0·079	0·15-0·20	—	0·005
ACTN/ACRN (tons/ton)	0·025	0·09	0·12	0·03-0·06	—	0·120
$H_2SO_4$100% (tons/ton)	—	—	0·20	—	—	—
Catalyst loss (grams/ton)	—	—	660 Bi_2O_3 440 MoO_3 (recoverable)	—	—	—
Electrical energy (kWh/ton)	—	—	800	—	—	—
Steam consumption (tons/ton)	—	—	14·5 with 6 from recovery	—	—	—
$(NH_4)_2SO_4$ produced (tons/ton)	0·400	—	0·375	—	—	—
Purity of ACRN	high	>99%	high	high	—	Crude 90% finished 98-99%

greater mechanical stress and abrasion which also results in greater catalyst loss. In Table 23 the losses for the OSW process are listed as 660 grams Bi_2O_3 and 440 grams MoO_3 per ton of acrylonitrile with the qualifying note that these losses may be recovered. No loss figures are available for the other processes. The catalyst life in the OSW process has been reported to be about one year and for the Distillers-Ungine process two years.

In general for all these processes a propylene conversion of 94-95 % per pass is or can be reached. The conversion of propylene to acrylonitrile (that is the yield of acrylonitrile with respect to propylene reacted) is in general only about 65-80 % (molar basis). The higher yields apply to fluidised bed processes and the lower ones to fixed bed processes.

As noted in Table 23 the percentage of secondary products produced (HCN, CH_3CN, $(NH_4)_2SO_4$) also varies appreciably with the type of process used. Thus the tons of HCN produced per ton of acrylonitrile varies from a low of 0·005 for the SNAM process to a high of 0·20 for the Distillers-Ungine process. Similarly for acetonitrile (CH_3CN) the range is from 0·025 tons per ton of acrylonitrile for the Montecatini process to 0·12 for the Sohio and SNAM processes.

The ammonium sulphate, included as a secondary product earlier, perhaps should be termed more correctly as a recovery product since it is formed by additions of sulphuric acid to the reactor effluent to adjust the

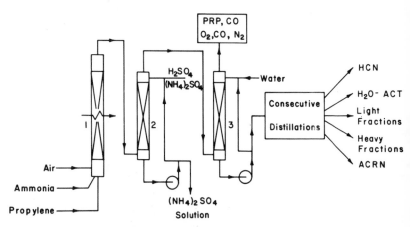

Fig. 59   Montecatini–Edison SPA process for manufacture of acrylonitrile.
1—fluidised bed catalytic reactor; 2—absorber for removal of unreacted ammonia; 3—absorber for removal of organic products; ACRN—acrylonitrile; ACT—acetonitrile; PRP—propylene.

pH of the ammonium sulphate solution which is used at this stage to remove the small amounts of unreacted ammonia from the product stream. This step is essential because in the presence of ammonia secondary reactions occur among the products which range from acrylonitrile, acetonitrile, hydrogen cyanide, acrylic aldehyde, acetaldehyde, acetone, propionitrile, and acrylic acid to carbon dioxide and carbon monoxide. It is the condensation reactions between the aldehydes, between the aldehydes and acids and between the aldehydes and the nitriles as well as with the ammonia that cause the problems. The polymerisation reactions may be sufficiently intensive and extensive (in the presence of the unreacted ammonia) to plug the pipes.

This treatment does not prevent nor remove the cyanohydrins which are formed by reactions between cyanohydric acid and the aldehydes and ketones in the product stream. The removal of these from the final product by distillation will be discussed briefly later.

THE MONTECATINI-EDISON AND BADGER CO. INC. PROCESS

The main characteristics of this process have already been summarised in Table 23.[414,439] A few further comments are in order to bring out additional technological features about it. A simplified flow diagram of this process is given in Figure 59. The ammoxidation reaction is effected in the fluidised bed reactor 7, at 420-460°C, about 2 atmospheres pressure and a contact time of 2 to 8 seconds. The reactor is cooled by heat exchange surfaces positioned within the fluidised bed, as stated earlier and the reaction products are cooled and scrubbed free of ammonia in the absorber-scrubber-2. The scrubbing liquor is primarily a solution of ammonium sulphate which is maintained slightly acidic by the addition of small quantities of sulphuric acid. The desired products are separated from the unreacted propylene, carbon dioxide oxygen and nitrogen by absorption in water in the absorber column 3. The aqueous solution of the organic products is then distilled consecutively in several columns to yield the final product acrylonitrile, a heavy fraction, a light fraction, an acetonitrile-water solution and hydrogen cyanide.

In pilot and semi-industrial scale operation, yields of 70-75 % based on propylene have been obtained. The purity of propylene employed commercially is 92-93 % and similar yields apply.

The catalyst employed in the Montecatini process is based on highly oxidised compounds of tellurium cerium and molybdenum deposited on a silica support in a ratio of 20-30 % oxides to 80-70 % silica. Under pilot plant evaluation the life of this catalyst was found to be of the order of 5000 hours, but it is dependent upon its mechanical and chemical stability which in turn is affected by the method of preparation.

The Sohio Process (Badger Co. Inc.) for the manufacture of acrylonitrile by ammoxidation of propylene in a catalytic fluidised bed reactor differs from the Montecatini fluidised bed process primarily in the nature of the catalyst used.[414,439-444] Initially the catalyst used was of the bismuth phosphomolybdate type but recently Sohio has replaced this with a new type named Catalyst 21.[445] This new catalyst contains uranium, which originates from atomic installations, and reportedly gives higher performances than the former, yielding higher conversions to acrylonitrile and increased amounts of secondary products, in particular hydrogen cyanide.

At the present time on a world basis about three quarters of the world-wide production (capacity) of acrylonitrile can be attributed to the Sohio process. The replacement of the old catalyst with the new put this process at a clear advantage above the others.[446]

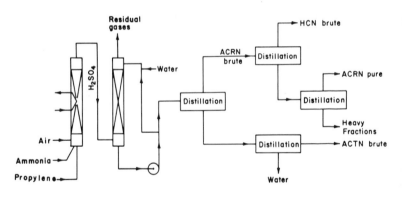

Fig. 60   Sohio-Badger Co. Inc. process for acrylonitrile manufacture.
1—catalytic fluidised bed reactor; 2—absorber for retention of products in water; ACRN—acrylonitrile ACTN—acetonitrile.

The flow diagram of the Sohio process is given in Figure 60. As specified in Table 22 the reaction temperature is somewhat greater (400-510°C) requiring a contact time of only a few seconds at the operating pressures of 0·5 to 2 atmospheres. As in the Montecatini process, the product stream is first scrubbed free of ammonia before the organic products are absorbed in water and then fractionated by distillation.

The Sohio process is characterised by the high degree of conversion of propylene to acrylonitrile per pass. Another feature about this process rests with the fact that sufficient amounts of hydrogen cyanide are co-produced to question the need of further expansion of the hydrogen cyanide industry by the classical process of reacting ammonia with hydrocarbons.[449] This is particularly true when it is recognised that the cost of hydrogen cyanide produced by this process is 50-60 % lower than by the classic routes.[448]

5.2 METHACRYLONITRILE

5.2.1 General Observations

Although methacrylonitrile is used to produce a variety of copolymers which have significant wide usage, it does not rank with acrylonitrile in importance in this regard. It is generally polymerised with acrylic acid, styrene, maleic anhydride and isoprene to produce lacquer protectors, organic glass and similar functional copolymers. Like acrylonitrile it can be produced by the ammoxidation reaction.

5.2.2 Manufacturing Processes

In relation to other available methods the ammoxidation of isobutylene to produce methacrylonitrile is superior. Although this process compares favourably in most respects with that for the ammoxidation of propylene at the present production levels, it does not have equal economics of scale and hence has higher unit production costs.

Other routes to methacrylonitrile include the alkylation of acetone—or acetaldehyde cyanohydrin with subsequent thermal decomposition of these esters according to the equation :[449-452]

$$CH_3-\underset{\underset{CH_3}{|}}{\overset{\overset{CN}{|}}{C}}-O-OR \xrightarrow[\text{HOOCH}]{430\text{-}450°C} CH_2=\underset{\underset{CH_3}{|}}{C}-CN$$

The overall yield is only about 54 % but the product produced is practically anhydrous and of high purity.

Intramolecular dehydrations of acetone cyanohydrin can be effected, also with the aid of phosphorus pentoxide[453] calcium oxide[458] and thionyl chloride either alone or in a mixture with an equivalent amount of pyridine.[454,455] The yield in this case is only 40-43%, but the product purity is even higher than by the thermal decomposition route.

Another method to produce methacrylonitrile is to react ammonia with isobutyl aldehyde.[456] The overall reaction can be given by the equation

$$(CH_3)_2CH-CH=O \xrightarrow[-H_2O]{NH_3, 550-600°C} CH_2=\underset{\underset{CH_3}{|}}{C}-CN$$

Finally this nitrile can be produced by reaction of ethylene oxide with acetonitrile in the presence of alumina at about 600°C.[457] In this case the following chemistry applies (yields reported are high):

$$CH_3CN + CH_2\underset{\underset{O}{\diagdown\diagup}}{-}CH_2 \xrightarrow{-H_2O} CH_2=\underset{\underset{CH_3}{|}}{C}-CN$$

5.2.3 Thermodynamic, Kinetic and Technological Considerations of the Ammoxidation of Isobutylene

The main reactions in this process may be summarised as follows:

(a) $(CH_3)_2C=CH_2 + NH_3 + 1·5O_2 \rightarrow CH_2=\underset{\underset{CH_3}{|}}{C}-CN + 3H_2O$

(b) $(CH_3)_2C=CH_2 + NH_3 + 4O_2 \rightarrow 4HCN + 8H_2O$

(c) $(CH_3)_2C=CH_2 + O_2 \rightarrow CH_2=\underset{\underset{CH_3}{|}}{C}-CHO + H_2O$

(d) $(CH_3)_2C=CH_2 + 4O_2 \rightarrow 4CO + 4H_2O$

(e) $(CH_3)_2C=CH_2 + 6O_2 \rightarrow 4CO_2 + 4H_2O$

(f) $4NH_3 + 5O_2 \rightarrow 4NO + 6H_2O$

(g) $4(CH_3)_2C=CH + 6NO \rightarrow 4CH_2=\underset{\underset{CH_3}{|}}{C}-CN + N_2 + H_2O$

It is evident that there is a close similarity between the chemistries of propylene and isobutylene ammoxidation processes. It is of interest here to use the available thermodynamic data as summarised in Table 24 to calculate the standard heat of reaction and equilibrium constant for the formation of methacrylonitrile from isobutylene, ammonia and air (reaction (a) above).

TABLE 24
Thermodynamic values for reactants and major products of methacrylonitrile manufacture

Chemical	Equation coefficients for $C_p = f(T)$				$H_{298°K}$ $(kcal/mol)$	$S_{298°K}$ (cal/mol)
	a	$b . 10^3$	$c . 10^6$	$d . 10^{-5}$		
isobutylene	5·331	60·24	−18·14	—	3·343	70·17
methacrylonitrile	6·570	44·74	−11·16	—	36·0	76·70
oxygen	8·643	0·202	—	−1·03	0	49·003
ammonia	6·187	7·887	−0·728	—	−11·04	46·01
water	7·170	2·56	—	0·08	−57·797	45·106
hydrogen	6·946	−0·199	0·480	—	0	31·21

The standard heat of reaction is found to be -123 kilocalories per mole, and the standard entropy change is 22·33 calories per mole degree. Hence using the free energy equation $G = H + TS$, the equilibrium constant can be evaluated as a function of temperature from the expression $\ln K_p = \Delta G_T/RT$. The following values show clearly that the equilibrium constant of ammoxidation of isobutylene to methacrylonitrile decreases with increasing temperature.

Temperature (°K)	298	498	698	898
$\log K_p$	85·341	49·105	33·633	25·058

In Figure 61 the yields of methacrylonitrile, carbon dioxide and hydrogen cyanide are graphed as a function of the molar ratio of ammonia to iso-butylene and of oxygen to isobutylene as well as a function of catalyst contact time.[459]

From Figure 61 it can be seen that an increase in the O_2 : IBTL ratio produces a significantly greater reduction in the yield of methacrylo-nitrile than a corresponding increase in the NH_3 : IBTL ratio. Indeed increasing the latter produces an initial increase in yield up to a ratio of about 3 beyond which the rate of yield reductions with further increases in

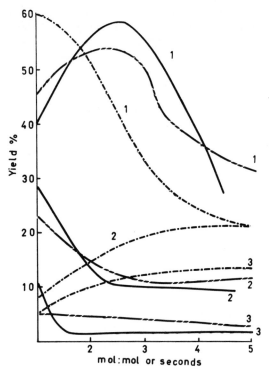

Fig. 61   Variations in yields of major products in isobutylene ammoxidation as a function of reactant composition and reaction time.
—— *for* $NH_3:IBTL$ *ratio effect* $IBTL:O_2:H_2O = 1:2:2\cdot5:1\text{-}3$; —·—·— *for* $O_2:IBTL$ *ratio effect* $IBTL:NH_3 = 1:2$; —————— *for contact time effect* $IBTL:NH_3:O_2:H_2O = 1:2:2\cdot5:1\text{-}3$.

the respective reactant ratios are comparable. The increase in the NH_3: IBTL ratio reduces the oxidation potential and hence diminishes the rates of the oxidation reactions (d) and (e). The reduction in yield of methacrylonitrile for NH_3:IBTL ratios greater than 3 may be ascribed to a reduction in the rate of the primary reaction (a), rather than to increases in secondary reactions. The effect of increases in contact time on yield of methacrylonitrile is similar to that of increasing NH_3:IBTL ratios. Presumably, again the maximum in yield can be ascribed to the rate of formation and concentration peaking of the intermediary compounds formed in primary reaction (a).

The yield of hydrogen cyanide decreases sharply initially as the NH_3: IBTL ratio is increased to about 1·5 beyond which further increases produce slight increases in yield. Increasing the contact time produces slight decreases in yield of hydrogen cyanide. Increases in the O_2:IBTL ratio up to about 3 increase the yield of hydrogen cyanide slightly beyond which further changes in the molar ratio have little to no effect on the yield of hydrogen cyanide.

Increases in the O_2:IBTL ratio have a similar effect on the yield of carbon dioxide as on the yield of hydrogen cyanide. On the other hand increases in the NH_3:IBTL ratio produces a sharp decrease in yield of CO_2 initially up to a value of about 2·5 beyond which further increases in the ratio produce little to no further change in yield of carbon dioxide. The effect of increases in contact time on the yield of carbon dioxide is difficult to reconcile; initially a fairly sharp reduction in yield occurs up to about 3 seconds beyond which a slight increase in yield of carbon dioxide occurs.

Thus, from the foregoing observations it appears that the conditions for an optimum yield of methacrylonitrile are a molar ratio of NH_3:IBTL of 2-3:1, a molar ratio of O_2:IBTL of 2-2·5:1 and a contact time of 2-3 seconds.

Further to the above considerations it should be noted here, also, that the total conversion of IBTL and the selectivity of this conversion to the nitrile are highly temperature dependent. Beyond a certain limit further increases in temperature produce considerable increases in total conversion but sharp decreases in selectivity. Thus at 420 and 440° the total conversions are about 80 and 92% and the selectivities about 50 and 44%, respectively. Modest increases in the contact time produce appreciable increases in total conversion and slight increases in the selectivity. Thus for contact times of 2 and 4 seconds the total conversions are 30-35% and 70-75% while the selectivities are 30-35% and about 45% respectively. The same sharp increase in IBTL total conversion and slight increase in selectivity appears with respect to increases in the molar ratio of O_2:IBTL.

Further with regard to the foregoing it has been reported that the total conversion of IBTL can reach 90-100% and the yield of methacrylonitrile 60-69% if a molar feed ratio of IBTL:NH_3:O_2:H_2O of 1:2-3:2-2·5:1-3 is employed with a contact time of about 3 seconds.[459]

Although acetonitrile has not been included in the graphs of Figure 61 or in the discussion thus far, it does appear in minor amounts in the ammoxidation process, also.

Studies on catalysts have confirmed that the catalyst systems utilised for the ammoxidation of propylene are equally effective for the ammoxidation of isobutylene. Among those tested and evaluated in this regard are systems based on molybdenum oxides[460] molybdenum and bismuth oxides and phosphomolybdates of bismuth[461] phosphates of manganese, molybdenum with and without tellurium oxide,[462,463] antimony oxide with ammonium molybdate and/or ammonium tungstate or vanadium and in oxides or iron and silicone oxides,[464-466] as well as oxides of tin and antimony.[467]

5.2.4 Technological Aspects

No special discussion on the technology for the manufacture of methacrylonitrile by the ammoxidation of isobutylene is warranted because it is so similar to that for the manufacture of acrylonitrile by the ammoxidation of propylene.

References

1. *Petrol Press Service*, **11**, 411-413 (1970)
2. *Chem. Age*, 2696, 13 (1971)
3. *Chimie Actualites*, 1405, 22 (1970)
4. *Chem. Ind.*, **21**, 10, 703 (1969)
5. *Gummi-Asbest-Kunststoffe*, **2**, 142 (1969)
6. Giudici, F. and Arcelli, B., *Parpinelli Informations Chim.*, 95, 153 (1971)
7. *Europe and Oil*, **8**, 10, 11 (1969)
8. *Materie Plast. Elast.*, **12**, 1604 (1969)
9. *Chem. Eng.*, **77**, 15, 36 (1970)
10. Sherwood, P. W., *Chimie et Ind.-Genie Chim.*, **102**, 3, 263 (1969)
11. *Chem. Eng.*, 238 (1953)
12. *Petrol Process.*, 82 (Nov. 1956)
13. *European Chem. News*, pp. 136 and 138 (1968)
14. *Hydroc. Process.*, **48**, 11, 197 and 198 (1969)
15. Rafikov, S. R., Suvarov, B. V., and Solomin, A. V., *Izvest. Akad. Nauk. Kasah.*, **1**, 58 (1957)
16. Dixon, J. K. and Longfeld, J. E., in: Emmet, P. H., *Catalysis*, New York, Reinhold Publ. Co., vol. 7, p. 185 (1960)
17. Margolis, Y. A. L., in: *Advances in Catalysis*, New York, Academic Press, vol. 14, p. 429 (1963)
18. Joffe, I. I. and Liubarskii, A. G., *Kin. i Kat.*, **3**, 261 (1962)
19. Sherwood, P. W., *Petrol. Proc.*, **11**, 82 (1956)
20. Benson, R. H. W. and Thorne, J. G. M., British Patent, 701,707 (1953)
21. Darby, J. R., U.S. Patent, 2,625,554 (1953)
22. Ryder, R. C., U.S. Patent, 2,885,409 (1959)
23. Darby, J. R., U.S. Patent, 2,674,582 (1954)
24. Darby, J. R., U.S. Patent, 2,624,744 (1953)
25. Amayasa, M., Watanabe, H., and Yoshikawa, A., Japanese Patent, 3522 (1958)
26. Dieibebbis, J. A., U.S. Patent, 3,005,831 (1961)
27. Skinner, W. A., and Tieszen, D., *Ind. Eng. Chem.*, **53**, 557 (1961)
28. Moldavskii, B. L. and Kernos, Y. D., *Kin. i Kat.*, **1**, 267 (1960)
29. Soc d'Electro Chim. d'Electro-Met. et Acier Electrique d'Ugine, Belgian Patent, 614,363 (1962)
30. Soc. d'Electro Chim. d'Electro-Met. et Acier Electrique d'Ugine, French Patent, 129,659 (1962)
31. *Hydroc. Process.*, **48**, 11, 199 (1969)
32. Seeboth, H., Buttner, W., and Rieche, A., *Brennst-Chemie*, **50**, 2, 47 (1969)

33. Moll Karl-Klaus and Fischer, E., *Chem. Technik*, **20**, 10, 600 (1968)
34. De Graaf, W. N., *S. Afr. Chem. Process*, p. 113 (1968)
35. Halcon International Inc., French Patent 1,462,595 (1966), US, 3,365,48: (1968)
36. Marchand, M. P., French Patent, 1,344,931 (1964)
37. California Res. Corp. British Patent, 946,916 (1964)
38. Toland, Jr., W. G., U.S. Patent, 3,246,028 (1966)
39. Hatch, L. F., *Hydroc. Process.*, **49**, 3, 101 (1970)
40. Prengle, H. W., Jr and Barona, N., *Hydroc. Process.*, **49**, 3, 106 (1970)
41. Halpern, J., *Disc. Faraday Soc.*, **46**, 7 (1968)
42. Waters, W. A., *Disc. of Faraday Soc.*, **46**, 158 (1968)
43. Pregaglia, G., Morelli, D., Conti, F., Georgio, G., and Ugo, R., *Disc. Faraday Soc.*, **46**, 110 (1968)
44. Aguillo, A., *Advan. Organometal. Chem.*, **5**, 321 (1967)
45. Clark, D., Hayden, P., and Smith, R. D., *Disc. Faraday Soc.*, **46**, 98 (1968)
46. Hartley, F. R., *Chem. Reviews*, **69** (6), 799 (1969)
47. Fenton, D. M., Olivier, K. L., and Biale, G., *Am. Chem. Soc. Petrol. Div. Preprint.*, **14** (4), C 77 (1969)
48. Okada, H., Noma, T., Katsuyama, Y., and Hashimoto, H., *Bull. Chem. Soc Japan*, **41**, 1395 (1968)
49. Bawn, C. E. H., *Disc. Faraday Soc.*, **14**, 181 (1953)
50. Walsh, A. D., *Trans. Faraday Soc.*, **42**, 269 (1946)
51. Helden, R. van, Bickel, A. F., and Kooyman, E. C., *Rev. Trav. Chim.*, **80**, 123? (1961)
52. Hay, A. S. and Blanchard, A. S., *Can. J. Chem.*, **43**, 1306 (1965)
53. Bateman, L., Gee, G., Morris, A. L., and Watson, W. F., *Disc. Faraday Soc.*, **10**, 250 (1951)
54. Bawn, C. E. H. and Williamson, J. B., *Trans. Faraday Soc.*, **47**, 721 (1951)
55. Bawn, C. E. H. and Williamson, J. B., *Trans. Faraday Soc.*, **47**, 735 (1951)
56. Carpenter, B. H., *Ind. Eng. Chem.*, *Proc. Des. Dev.*, **4**, 105 (1965)
57. Burney, D. E., Weisemann, G. H., and Fragan, N., *Pet. Ref.*, **38**, 6, 186 (1959)
58. Fortuin, J. P., Waale, M. J., and Oosten, R. P., *Pet Ref.*, **38**, 6, 189 (1959)
59. Ravens, D. A. S., *Trans. Faraday Soc.*, **55**, 1768 (1959)
60. Towle, P. H. and Baldwin, R. H., *Hydroc. Process.*, **43**, 11, 149 (1964)
61. Bart, E. V., Joffe, I. I., Nikonova, E. K., Piskunova, E. P., and Stepikov, A. V., *Neftekhimiya*, **9**, 6, 873 (1969)
62. Perozici, D. I., *Chim. Prom.*, 1, 10 (1969)
63. Krekeler, H. and Schmitz, H., *Chem. Ing. Techn.*, 785 (1968)
64. Ramirez, R., *Chem. Eng.*, 75, 17, 94 (1968)
65. Schwerdtel, W., *Chem. and Ind.*, p. 1559 (1968) (English)
66. Krekeler, H. and Kroning, W., Seventh World Petroleum Congress, vol. 5, *Petro Chemistry*, London, Elsevier Publishing Co. Ltd., p. 41 (1967)
67. Schaeffer, W. D., U.S. Patent 3,277,158 (1966).
68. Farbwerke Hoest, A. G., German Patent, 1,216,290 (1966)
69. E. I. DuPont de Nemaurs and Co., British Patent, 1,083,959 (1967)
70. Schaeffer, W. D. and Olivier, K. L., U.S. Patent, 3,346,626 (1967)
71. *Hydroc. Process.*, **48**, 11, 247 (1969)
72. Smidt, T., *Angew. Chem.*, **71**, No. 5, 176 (1959)

73. Schmidt, J., Hafner, W., Sedlmeier, J., Jira, R., and Ruttinger, R., U.S. Patent, 3,131,223 (1964)
74. Smidt, J., *Chem. Ind.*, London, p. 54 (1962)
75. Riemenschneider, W., Dialer, K., Probst, O., and Bander, O. E., U.S. Patent, 3,201,905 (1967)
76. *Hydroc. Process*, **48**, 11, 137 (1969)
77. *Information Chemie*, **58**, 35 (1968)
78. Ricmenschneider, W. and Dialer, K., U.S. Patent, 3,076,032 (1963), 3,118,001 (1964)
79. Hörnig, L., Paszthory, E., and Wimmer, R., U.S. Patent, 3,149,167 (1964)
80. Smidt, J., *Chem. Eng. News*, **41**, 50 (1963)
81. *Hydroc Process.*, **48**, 11, 140 (1969)
82. Andreas, F. and Gröbe, K., *Propylen Chemie*, Akademie Verlag, Berlin, p. 174, 175 (1969)
83. *Hydroc. Process.*, **48**, 11, 204 (1969)
84. Henry, P. M., *J. Am. Chem. Soc.*, **88**, 1595 (1966)
85. Banstead, A. E., Stanley, H. M., and Tuerk, K. H. W., U.S. Patent, 2,514,041 (1950)
86. Benson, G., *Chem. Met. Eng.*, **47**, 3, 150 (1940)
87. McFarlane, S. B., U.S. Patent, 2,491,572 (1949)
88. Alheritier, L., Biarnais, P., and Sitand, G., U.S. Patent, 3,258,483 (1966)
89. Hester, A. S. and Himmler, K., *Ind. Eng. Chem.*, **51**, 12 (1959), 1429 (1959)
90. Riesser, G. H. and Smith, R. F., U.S. Patent, 3,281,462 (1966)
91. Emanuel, N. M., *Dokl. Akad. Nauk. SSSR*, **110**, 245 (1956)
92. Johnson, P. C. and Masher, D. R., U.S. Patent, 3,282,994 (1966)
93. Sennevald, K., Vegt, W., Erpenback, H., and Joest, H. I., U.S. Patent, 3,335,160 (1967)
94. Weymouth, F. J. and Millidge, A. F., *Chem. Ind.*, **22**, 887 (1966)
95. Schaeffer, W. D. and Olivier, K. L., U.S. Patent, 3,293,292 (1966)
96. Furman, M. S., Shestakowa, A. D., Arest Yakubovich, I. L., and Lyubitsyna, N. A., *Dokl. Akad. Nauk. SSSR*, **124**, 1083 (1959)
97. Rouchoud, J. and Lutete, B., *Ind. Eng. Chem. Prod. Res. Dev.*, **7**, 4, 266 (1968)
98. Asinger, F., *Paraffins Chemistry and Technology*, Pergamon Press, N.Y. (1968)
99. Perchenko, A. A. and Marchenko, M. A., *Khimiya i tekhnologiya topliv i masei*, **15**, No. 4, 30 (1970)
100. Hull, D. C., U.S. Patent, 2,673,217 (1952)
101. Banker, R. S. and Saffer, A., U.S. Patent, 2,963,509 (1969)
102. Rhodes, M. C., U.S. Patent, 3,089,970 (1963)
103. SNIA Viscosa, Italian Patent, 803,424 (1968)
104. Donati, I., Sioli, G., and Taverna, M., *La Chim. e L'Ind.*, **50**, 9, 997 (1968)
105. Taverna, M., *La Rivista dei Comb.* **XXII**, 4, 203 (1968)
106. Weaver, C. W., U.S. Patent, 3,235,588; Fragen, M., 3,210,416; Hundley, J. G., 3,187,038; Ashcraft, J. O., Winstrom, L. O., Duggan, R. J. and Park, J. C., 3,321,382 (1966)
107. Giborowski, S., German Patent (R.F.G.), 1,158,052; (1963) 1,189,977 (1966)
108. Pavlova, P. S., Olevskii, V. M., and Popow, D. M., *Him. Prom.*, **1**, 2 (1969)

109. Kaeding, W. W., Lindblom, R. O., and Temple, R. G., U.S. Patent, 2,727,926 (1965)
110. Taplin III, W. H., U.S. Patent, 2,954,407 (1960)
111. Castabello, D., Forni, A., and Ramello, P., French Patent, 1,339,394 (1963)
112. *Hydroc. Process.*, **48**, 11, 215 (1969)
113. Kaeding, W. W., Lindblom, R. O., and Temple, R. G., *Ind. Eng. Chem.*, **53**, 805 (1961)
114. Kaeding, W. W., *Hydroc. Process*, **43**, 11, 173 (1964)
115. Kaeding, W. W., Lindblom, R. O., Temple, R. G., and Mahon, H. I., *Ind. Eng. Chem., Proc. Res. Dev.*, **4**, 97 (1965)
116. Gardner, J. M., Tewksbury, C. I., U.S. Patent, 2,784,202 (1957)
117. Aries, R. S., U.S. Patent, 3,071,601 (1963)
118. Gash, V. W., U.S. Patent, 3,275,662 (1966)
119. Sharp, D. B., U.S. Patent, 3,232,957 (1966)
120. Shinogi and Co. Ltd, Japanese Patent, 69,26,282 (1969)
121. Guy, L. and Hubert, L., German Patent, 1,904,573 (1969)
122. Nobuyoshi, S., Masao, O., and Mischitoshi, K., Japanese Patent, 69,12,128 (1969)
123. Badishe Aniline and Soda Fabrik, A.G., British Patent, 1,147,285 (1969)
124. Badische Aniline and Soda Fabrik, A.G., British Patent, 1,123,514 (1968)
125. Badishe Aniline and Soda Fabrik A.G., French Patent, 1,532,005 (1968)
126. Schulz, J. G. D. and Whitaker, A. C., U.S. Patent, 3,390,174 (1968)
127. Wiener, M. V., British Patent, 1,107,427 (1968)
128. Jacque, B., Bernard, B., Michael, G., and Raymond, J., French Patent, 1,512,534 (1968)
129. Lidov, R. E., U.S. Patent, 3,361,806 (1968)
130. Rouchoud, J. and Chantrain, B., *Bull. Soc. Chim. Fr.*, **4**, 1329 (1968)
131. Buxbaum, H. L., *Justus Liebig Ann. Chem.*, **706**, 81 (1967)
132. Steeman, J. W. M., Kaarsemaker, S., and Hoftyzer, P. J., *Chem. Eng. Sci.*, **14**, 1961, 139 (1961)
133. *Oil and Gas Journal*, **67**, 50, Dec. 15, 74 (1969)
134. Spielman, M., *Advan. in Chem. Eng. Journ.*, **4**, 496 (1964)
135. *Hydroc. Process.*, **8**, 11, 170 (1969)
136. Friedrich, B., Heinz, V., and Zurtz, P., French Patent, 1,527,392 (1968)
137. Viskers-Zimmer, A. G., British Patent, 1,145,815 (1969)
138. Lindsay, A. F., *Chem. Eng. Sci.*, Suppl. 1 to vol. 3, 78 (1954)
139. Trubnikova, V. I., *Him. Prom.*, **46**, 1, 12 (1970)
140. Flemming, W. and Speer, W., U.S. Patent, 2,005,183 (1935)
141. Amend, W. J., U.S. Patent, 2,316,543 (1943)
142. Parlant, C., Serre de Roch, I. and Balaceau, J. C., *Bull. Soc. Chim. France*, **11**, 2452 (1963)
143. Lacey, R. N. and Allison, K., U.S. Patent, 3,258,491 (1966)
144. Saffer, A. and Barker, R. S., U.S. Patent, 2,833,816 (1958)
145. Burney, D. E., Weisman, G. H., and Fragen, N., *Pet. Ref.*, **38**, 6, 186 (1959)
146. Towle, P. H. and Baldwin, R. N., *Hydroc. Process.*, **43**, 11, 149 (1964)
147. Spiller, C. A. and Malo, R. V., U.S. Patent, 3,029,278 (1962)
148. Yoshimura, T., *Chem. Eng.*, **74**, May 5, 78 (1969)
149. *Hydroc. Process.*, **48**, 11, 240 (1969)

150. Digurov, N. G., *Himia l Him. tehn.*, **13**, 3, 407 (1970)
151. Nicolescu, I. V., Angelescu, E., Parauseanu, V., Mihaita, St., Tudor, G., and Hogea, I., 39-Congr. Intern. Chim. Ind. Buccuresti-Romania, vol. 2, 7/125, 7-11 Sept. 1970
152. Ivanov, A. M., Chervinskii, K. A., and Baranova, E. I., *Neftehimiya*, **9**, 6, 892 (1969)
153. Khcheyan, Kh. E., Parlicev, A. T., Arbitman, S. M., and Kuricheva, L. N., *Him. Prom.*, **6**, 6 (1962)
154. Landau, R., *Ind. Chemist*, **33**, 344, 615 (1957)
155. Null, H. R., Bowe, L. E., and Binning, R. C., U.S. Patent, 3,335,179 (1967)
156. Krönig, W., Seventh World Petroleum Congress, vol. 5, *Petrochemistry*, London, Elsevier Publishing Co. Ltd, p. 59 (1967)
157. Morrison, J. J., *Oil and Gas J.*, **68**, 47, 57 (1970)
158. Roy, G., *Chimie et Ind.*, *Genie Chimique*, 103, 7, p. 777 (1970) French Patent, 1,506,296 (1966)
159. *Hydroc. Process.*, **48**, 11, 214 (1969)
160. Landau, R., Brown, D., Russell, J. L., and Kollai, J., Seventh World Petroleum Congress, vol. 5, *Petrochemistry*, London, Elsevier Publishing Co. Ltd, p. 67 (1967)
161. Allen, I. A., *Chem. and Ind.*, **30**, 1225 (1963)
162. Calderbank, P. H., *Ind. Chemist*, **28**, 291 (1952)
163. Mars, P. and van Krevelen, D. W., *Chem. Eng. Sci.*, **3**, Spec. Suppl. (1954)
164. Emmett, P. H., *Catalysis*, vol. 7, New York, Reinhold Publ. Corp., p. 198 (1960)
165. Ioffe, I. I. and Sherman, Yu, G., *J. Fiz. Khim.*, **29**, 692 (1955)
166. Pinchbeck, P. H., *Chem. Eng. Sci.*, **6**, 106 (1957)
167. D'Alessandro, A. F. and Farkas, A. I., *Colloid Sci.*, **11**, 653 (1956)
168. Roiter, V. A., Ushakova, V. P., Korneichuk, G. P., and Shorbilina, T. G., *Kin. i Kat.*, **2**, 94 (1961)
169. Kasatkina, L. A. and Boreskov, G. K., *J. Fiz. Khim.*, **29**, 455 (1959)
170. Emmett, P. H., *Catalysis*, vol. 7, New York, Reinhold Publ. Corp. (1960)
171. Roiter, V. A., Korneichuk, G. P., Ushavova, V. P., and Stukanouskayoc, N. A. *Cataliticeskoie Okislenie naftalina, Kiev, Izv. Akad. Nauk SSSR* (1963)
172. Bhattacharyya, S. and Guloti, F., *Ind. Eng. Chem.*, **50**, 1719 (1958)
173. Friederichsen, W., *Chem. Ing. Techn.*, **41**, 17, 967 (1969)
174. Wojtowicz, T., Ghzesik, A., Czanota, T., and Ligeza, A., *Przem. Chem.*, **44**, 200 (1965); *idem*, **46**, 578, 663 (1967)
175. Klema, F., *Chem. Ztg.-Chem. App.*, **91**, 21, 815 (1967)
176. Guccione, E., *Chem. Eng.*, **72**, 12, 132 (1965)
177. FIAT 649
178. Toland, Jr., W. G., U.S. Patent, 2,574,511 (1950)
179. Levine, I. E., U.S. Patent, 2,521,466 (1950)
180. Marek, L. F., *Ind. Eng. Chem.*, **41**, 1894 (1949)
181. Betts, W. D., *Ind. Chemist.*, **39**, 6, 302 (1963)
182. *Brit. Chem. Eng.*, **2**, p. 8 (1957)
183. Litvinenko, M. I., *J. Obsc. Khim.*, **1**, 81 (1961)
184. Ackerman, P., *Pet. Ref.*, **44**, 11, 291 (1965)
185. *Hydroc. Process.*, **48**, 11, 217, 218, 219 (1969)

186. Gurevich, D. A., *Khim. Prom.*, **32**, 9, 647 (1967)
187. *Europ. Chem. News*, **12**, 295, 40 (1967)
188. Graham, J., Europ. *Chem. News*-Large Plant Suppl., 27 (1966)
189. Duckworth, R. A., *Chem. Process. Eng.*, **1**, 69 (1969)
190. Schach, A. and Krings, W., German Patent 1,109,658 (1960)
191. Waterman, W. W., U.S. Patent, 2,901,441 (1959)
192. T'eng-ko Jen: *Chemistry* (Taipeh), (1959), p. 7; *Chem. Abstr.*, **54**, 1960, p. 5606
193. Endler, M., Italian Patent, 553,705 (1957)
194. Herman, M., Endler, A. E., and Bulgarelli, E., U.S. Patent, 2,825,701 (1958)
195. Singus, H., U.S. Patent, 2,985,668 (1961)
196. Metzger, F. J., U.S. Patent, 2,805,229 (1957)
197. Tsutsumi, S., Japanese Patent, 3,859 (1957); *Chem. Abstr.*, **52**, 1958, p. 5476 c.
198. Sakuyama, S. and Kiguchi, I., Japanese Patent, 9,307 (1957); *Chem. Abstr.*, **52**, 1958, p. 15589 c
199. Egbert, R. B., U.S. Patent, 2,764,598 (1957)
200. Chemiche Werke Huls. G. m.b.H. British Patent, 755,218 (1956)
201. Shiino, K., *Reports, Govt. Chem. Ind. Res. Inst.*, Tokyo, **50**, 70 (1950); *Chem. Abstr.*, **50**, 1956, p. 10073 f
202. Egbert, R. B., U.S. Patent, 2,693,474 (1954)
203. Schach, A. and Kings, W., British Patent, 892,381 (1962)
204. Hoffman, E., Timbrel Ltd, Australian Patent, 142,908 (1951)
205. Badishe Aniline & Soda Fabric, German Patent, 825,835 (1951)
206. Lichtenwalter, M., Peterson, E. and Sacken, D. K., U.S. Patent 2,769,016 (1956)
207. Shoenberg, M., Wirtz, R., Dialer, K., and Kalhammer, F., German Patent, 1,108,197 (1958)
208. Block, H. S., U.S. Patent, 3,054,807 (1962)
209. Nault, G., *Ind. Eng. Chem. Proc. Des. Dev.*, **1**, 285 (1962)
210. Orzechowski, A. and MacCormack, K. E., *Can. J. Chem.*, **32**, 388 (1954)
211. Kurilenko, A. I., Kilkova, N. V., Rylokova, N. A., and Temkin, N. I., *J. Fiz. Khim.*, **32**, 797, 1043 (1958)
212. Boreskov, G. K., *Kin. i Kat.*, **3**, 214 (1962)
213. McBee, E. T., Hass, H. B., and Wiseman, P. A., *Ind. Eng. Chem.*, **37**, 432 (1945)
214. Wan, S., *Ind. Eng. Chem.*, 45, 234 (1953)
215. Janda, J., *Chem. prumysl.*, 14/39, 469 (1964)
216. Law, G. M. and Chitwood, M. C., British Patent, 518,823 (1940)
217. Landau, R., U.S. Patent, 2,785,186 (1957)
218. Chem. Patents Inc., British Patent, 742,489 (1955)
219. Drummond, V. D., Gould, M. L., and Katzen, R., U.S. Patent, 2,752,363 (1956)
220. N.V. de Bataafsche Petr. Maatschappij, British Patent, 704,227 (1954)
221. Ostrovski, V. E., Kulkova, N. V., Lopatin, V. L., and Temkin, M. I., *Kin. i Kat.*, **3**, 189 (1962)
222. *Petrol Ref.*, **42**, 11, 171, 173 (1963)
223. Goddu, R. F. and Delker, D. A., *Anal. Chem.*, **30**, 2013 (1958)

224. Amberg, C. H., Echigoya, E., and Kiławic, D., *Can. J. Chem.*, **37**, 708 (1959)
225. *Hydroc. Process.*, **48**, 11, 178 (1969); *idem*, **32**, 9, 146 (1953)
226. *Hydroc. Process.*, **46**, 11, 175 (1967); *idem*, **48**, 11, 179 (1969)
227. Morrison, J., *Oil and Gas J.*, **68**, 47, 61 (1970)
228. Kulkova, N. V., *Him. Pro.*, **44**, 9, 16 (1968)
229. Emanuel, N. M., *J. Vsesoiuznugo Khim. Obsc. im Mendeleeva*, **14**, 3, 248 (1969)
230. *Information Chimie*, **63**, 99 (1968)
231. *Chemische Industrie*, **5**, 342 (1968)
232. *Ekonomida Him. Prom.*, **2**, 25 (1967)
233. *Chem. Week*, **98**, 10, 27 and 30 (1966)
234. Clement, J., *L'Ind. Chimique*, **616**, 455 (1968)
235. *Hydroc. Process.*, **48**, 11, 146, 147 (1969)
236. *Chem. Processes in Europe*, p. 61 (1966)
237. Isaev, O. V., Margolis, L. Ya, and Sazonova, J. S., *Dokl. Akad. Nauk SSSR.*, **129**, 141 (1959)
238. Golovina, O. A., Isaeve, V. I., and Sacharov, M. M., *Dokl. Akad. Nauk SSSR*, **142**, 619 (1962)
239. Sixma, F. L., Duynstee, E. F. J., and Hennekens, J. L. J. P., *Rec. Trav. Chim. Pays Bas*, **82**, 901 (1963)
240. Voge, H. H., Wagner, C. D., and Stevenson, D. P., *J. Catalysis*, **2**, 58 (1963)
241. Hearne, G. H. and Adams, M. L., U.S. Patent, 2,451,485 (1948), *idem*, 2,486,842 (1949)
242. Schmidt, P., U.S. Patent, 2,848,475 (1958)
243. Hackmann, Johannes Th., Danish Patent, 69,025 (1951)
244. Hadley, D. J. and Goodings, E. P., British Patent, 658,179 (1951)
245. Hadley, D. J., U.S. Patent, 2,670,380 (1954)
246. Hamada, R. H., Japanese Patent, 7,513 (1958)
247. Barclay, T. L., Bethell, J. R., Bhean, J. B., Hadley, D. J., Jenkins, R. M., Steward, D. G., and Wood, B., British Patent, 864,666 (1961)
248. Barclay, J. L., Hadley, D. J., and Stewart, D. G., British Patent, 873,712 (1962)
249. Stamicarbon, N.V., Dutch Patent, 106,994 (1963); French Patent, 1,316,876 (1962)
250. Imperial Chemicals Ind. Ltd, French Patent, 1,380,884 (1964)
251. Veatch, F., Callahan, J. L., Milberg, E. C. and Foreman, A. W., Proceedings 2nd International Congress on Catalysis, Paris, 1960, Vol. 2, pp. 2647–58 (see *Chem. Abs.*, **55**, 24547 e)
252. Adams, C. R. and Jennings, T. J., *J. Catalysis*, **2**, 63 (1963)
253. Brown, C. and Newman, F. C., French Patent, 1,399,498 (1964)
254. Deutsche Gald-wnd Silber Scheideanstalt vorm Roessler, French Patent, 1,392,718 (1964)
255. Ruhrchemie, A. G., French Patent, 1,366,580 (1964)
256. Brown, C. J. and Millage, A. F., British Patent, 953,763 (1962)
257. Bethell, J. R., Hadley, D. J., Gasson, E. J., and Neale, R. F., British Patent, 903,034 (1962)
258. Idol, J. D., Jr., Callahan, T. L., and Foreman, R. W., U.S. Patent, 2,881,212 (1959)

259. Voge, H. H., Morgan, C. Z., Sachtler, M. H., and Ryland, L. B., Belgian Patent, 614,404 (1962)
260. Imperial Chemical Ind. Ltd, French Patent, 1,364,810 (1964)
261. Wakabayashi, K., *Bull. Chem. Soc. Japan*, **41**, 11, 2776 (1968)
262. BASF Dutch Patent, 69-08212 (Aplication May 29, 1969), *European Chem. News*, **17**, 428, 55 (1970)
263. Howlett, J. and Lamburd, C. A., British Patent, 687,852 (1949)
264. N.V. de Bataafsch, Petr. Madtschappij, British Patent, 668,340 (1952)
265. British Celanese Ltd, British Patent, 655,902 (1951)
266. Klein, D. Y. and Kvalnes, D. E., U.S. Patent, 2,476,307 (1949)
267. *Ind. Eng. Chem.*, **62**, 9, 63 (1970)
268. *Him. Prom. za rubejom*, **12**, 47 (1969)
269. Nakatani, Haruo, *Japan Chem. Quat.*, **IV-III**, 4, 3, 50 (1968)
270. Nakatani, Haruo, *Hydroc. Process.*, **48**, 5, 152 (1969)
271. *Hydroc. Process.*, **48**, 11, 145 (1969)
272. *Europ. Chem. News*, **16**, 391, 19 (1969)
273. Miller, S. A. and Donaldson, J. W., *Chem. and Process Eng.*, **48**, 12, 37 (1967)
274. Stobaugh, R. B., *Petro. Manuf. Mark. Guide*, vol. I; *Arom. and Deriv.*, p. 64 (1966)
275. Ohlinger, H. and Stadelmann, S., *Chem. Ing. Technik*, **37**, 361 (1965)
276. Boundy, R. H. and Boyer, R. F., *Styrene*, New York, Reinhold Publ. Corp. (1952)
277. Whitby, G. S., *Synthetic Rubber*, New York, J. Wiley (1954)
278. Carra, S., *La Chim. e l'Ind.*, **45**, 949 (1963)
279. Wenner, R. R. and Dybdal, C. E., *Chem. Eng. Progr.*, **44**, 275 (1948)
280. Balandin, A. A., Bogdanava, O. K., and Belomestnykh, J. P., *Dokl. Akad. Nauk. SSSR.*, **138**, 3, 595 (1961)
281. Berdutin, A. Y., Terekhim, R. M., and Yukelson, I. I., *Him. Prom.*, **45**(9), 662-5 (1969)
282. Weber, H. W., Jr., U.S. Patent, 3,299,155 (1967)
283. Bajars, L., U.S. Patent, 3,308,187 (1967)
284. Bajars, L., U.S. Patent, 3,308,188 (1967)
285. Bajars, L., U.S. Patent, 3,308,198 (1967)
286. Olson, D. H., U.S. Patent, 3,406,219 (1968)
287. Esso Research and Eng. Co., French Patent, 1,555,994 (1968)
288. Adams, C. R., *Ind. Eng. Chem.*, **61**, 6, 157 (1969)
289. Csomontany, G., Balmez, G., and Netta, M., 39th Congr. Internat. Chim. Ind., Vol. 2, pp. 8-19, 7-11 Sept. Bucuresti-Romania (1970)
290. Berak, J. M., Scharf, E., Wozniakiewicz-Baj, M., Jelowek, Z., Lazewski, T., and Kehl, B., *Przem. Chem.*, **48**, 12, 725 (1969)
291. Froment, G., *Ind. Chim. Belge*, **27**, 1041 (1962)
292. Tiuramev, I., Bushin, A. N., Mckailov, R. K., and Saryskeva, E. A., *Fiz. Chim.*, **31**, 93 (1957)
293. Yuichi Ozawa and Bischoff, K. B., *Ind. Eng. Chem. Process Design and Devel.*, **7**, 1, p. 72 (1968)
294. Coriciuc, C., Csomontany, G., and Popa, Sp., 39th Congr. Internat. Chim. Ind., 7-11 Sept., Bucuresti-Romania, vol. 2, pp. 8-14 (1970)

295. Hohlov, V. A., Joblonski, G. S., and Vepseva, S. V., *Him. Prom.*, **10**, 726, 6 (1968)
296. Froment, G. and Bischoff, K. B., *Chem. Eng. Sci.*, **16**, 189 (1961)
297. Voorhies, A., Jr., *Ind. Eng. Chem.*, **37**, 318 (1945)
298. Kubica Benedkt, *Przem. Chem.*, **47**, 12, 753 (1968)
299. Soderquist, F. J., Boyce, H. D., and Kline, P. E., U.S. Patent, 3,435,086 (1969)
300. Ward, D. J., U.S. Patent, 3,409,689 (1968)
301. McMahon, J. F., U.S. Patent, 3,371,125 (1968)
302. McMahon, F., French Patent, 1,500,632 (1967)
303. Weber, H. W., Jr., U.S. Patent, 3,299,155 (1967)
304. Downs, R. O. and Franz, R. A., U.S. Patent, 3,441,457 (1969)
305. Bogdanova, O. K., Balandin, A. A., and Belomestingkh, I. P., *Dokl. Akad. Nauk SSSR.*, **146**, No. 6, p. 1327 (1962); *ibid* 132, No. 2, p. 343 (1960)
306. Graves, G. D., U.S. Patent, 2,036,410 (1936)
307. Stanley, H. M. and Salt, F. E., U.S. Patent, 2,342,980 (1944)
308. Balandin, A. A., Marukyan, G. M., and Seimovich, R. G., *Compt. rend. acad. sci. URSS*, **41**, 67-9 (1943) (Engl.)
309. *Hydroc. Process.*, **48**, 11, 234, 235 (1969)
310. Abraham, D. C., German Patent, 1,806,255 (1969)
311. Victor, V., French Patent, 1,540,367 (1968)
312. Henry, J. P. and Wilkinson, L. A., U.S. Patent, 3,326,996 (1967)
313. Askerov, A. K., *Azerbaid, nefteanoe hoz.*, **48**, 11, 35 (1969)
314. Schwachula, G. and Wolf, F., *Chem. Technik.*, **22**, 3, 163 (1970)
315. Kearby, K. K., *The Chem. of Pet. Hydroc.*, Reinholds Pub., vol. 2, p. 221 (1955)
316. Liubarskii, G. D., *Uspehi Himii*, **27**, 316 (1958)
317. Balandin, A. A., *Advan. Catalysis*, **10**, 96 (1958)
318. Tiurisev, I. Ia., *Uspehi Himii*, **35**, 121 (1966)
319. Asinger, F., *Chemie und Technologie der Monoolefine*, Berlin Akademie-Verlag (1957)
320. Kearby, K. K. and Emmet, P. H., *Catalysis*, New York, Reinhold Publ. Corp., vol. III, p. 453 (1965)
321. Sanford, R. A. and Patinkin, S. H., U.S. Patent, 2,958,715 (1960)
322. Foster, R. L., Wunderlich, D. K., Patinkim, S. H., and Stanford, R. A., *Pet. Ref.*, **39**, 11, 229 (1960)
323. Di Giacomo, A. A., Maerker, J. B., and Schall, J. W., *Chem. Eng. Progr.*, **57**, 5, 40 (1961)
324. Heineman, H., Millikey, T. H., Jr., and Stevenson, D. H., French Patent, 1,203,754 (1958)
325. Maerker, J. B., U.S. Patent, 2,946,831 (1960)
326. *Chem. Week*, **87**, 18, 39 (1960)
327. Shell Internat. Res. Macetschappij, N.V., British Patent, 867,296 (1961)
328. Katsen, R., *Neftepererabotka i neftehimia*, **10**, 47 (1964)
329. Begley, W. I., *Pet. Ref.*, **44**, 7, 149 (1965)
330. Skinner, I. L. and Slipcevich, C. M., *Ind. Eng. Chem.*, **2**, 161 (1963)
331. Beckberger, L. M. and Watson, K. M., *Chem. Eng. Progr.*, **44**, 229 (1948)
332. Weller, S. W. and Volts, S. E., *Z. phys. Chem.*, **5**, 100 (1955)

333. Trifiro, F. and Pasquon, I., *La Chim. e L'Ind.*, **52**, 3, 228 (1970)
334. Kolubihin, V. A., Emelianova, E. N., Valtev, V. N., and Miasedov, M. I., *Him. Prom.*, **11**, 825, 25 (1968)
335. Ivasina, V. S., Buyanov, R. A., Ostan'kovich, A. A., Olen'kova, I. A., Kotel'nikov, G. A., Kefeli, L. M., and Stupnikova, L. V., *Kin. i Kat.*, **11**, 1, 160 (1970)
336. Bennet, I. A. R., *Chem. and Ind.*, **14**, 410 (1961)
337. Harbour, R. J., *Brennstoff Chemie*, **12**, 389 (1959)
338. Hornaday, G. F., Ferrell, F. M., and Mills, G. A., *Advan. Petrol. Chem.*, **4**, New York, Interscience Publishers, p. 451 (1961)
339. Shendrik, M. N., Bereskov, G. K., and Kirilyuk, L. V., *Kin. i Kat.*, **10**, 2, 280 (1969)
340. Timoshenko, V. I., Buyanov, R. A., and Proskii, O. I., *Kin. i Kat.*, **10**, 3, 681 (1969)
341. Hapel, J., Blank, H., and Hamill, T. D., *Ind. Eng. Chem., Fund.*, **5**, 3, 289 (1966)
342. Carra, S., Formi, L., and Vintani, C., *J. Catalysis*, **IX**, 2, 154 (1967)
343. Serban, Gh., Goidea, D., Andrei, G., Ceafalan, N., and Goidea, N., 39th Congr. Internat., Chim. Ind., Bucuresti-Romania, vol. 2, pp. 8-110, 7-11 Sept. (1970)
344. Annenkova, I. B., Alkhazov, T. G., and Belen'kii, M. S., *Kin. i Kat.*, **10**, 6, 1305 (1969)
345. Jones, J. H., Danbert, T. E., Feuske, M. R., Sandy, C. W., and Lou, P. J., *Ind. Eng. Chem., Proc. Design and Devel.*, 9, 1, 127 (1970)
346. Winnick, C. N., French Patent, 1,244,499 (1959)
347. Mihailov, R. K., *Him. Prom.*, **46**, 1, 3 (1970)
348. Bogdanova, O. K., Shcheglove, A. P., and Baladin, A. A., *Neftehimia*, **2**, 442 (1962)
349. Bajars, L., U.S. Patent, 3,377,403 (1968), 3,374,283 (1968)
350. Adelson, S. V. and Nikonova, M. M., *Tr. Mosk. Inst. Neftehim. Gazov Prom., USSR*, **72**, 9 (1967)
351. Tiurgaev, I., *Neftehimia*, **6**, 1, 71 (1966)
352. Felterly, L. C., U.S. Patent, 3,391,213 (1968)
353. Serebreakov, B. R., Dalin, M. A., Gusman, T. Ya, Kockharov, V. G., Galanternik, R. E., and Muslin-Zade, Z. M., *Neftepererab. Neftehimii, USSR*, **2**, 28 (1968)
354. Adelson, S. V., Nikonova, M. M., and Nikonov, V. I., *Himii i topliv i masel*, **15**, 1, 8 (1970)
355. Sekusova, X. Z., Izv. vis. uceh. zaved., *Himiia i Himic. tehn.*, **13**, 1, 102 (1970)
356. Dadaesev, B. A., Sonydzhanov, A. A., and Dogzhi-Kasunov, V. S., *Azeraidj. nefteance hoz.*, **50**, 1, 36 (1970)
357. Hodakov, I. S., Minauev, H. M., and Sterlingnov, O. L., *Dokl. Akad. Nauk SSSR*, **165**, No. 2, 344 (1965).
358. Breiman, M. I., Liukumovich, A. G., Rutman, G. I., Penomareva, A. P., and Ponomarenko, V. I., *Nefteperabotka Neftehimia*, **2**, 24 (1965)
359. Yukelson, I. I. and Boguslavskii, E. A., *Neftehimiia*, **4**, 834 (1964)
360. Kolobikhin, A. V. and Emelyanova, E. N., *Neftehimiia*, **4**, 829 (1964)
361. Whitby, G. S., *Synthetic Rubber*, New York, John Wiley (1964)

362. Kolobikhin, V. A., Soboliv, V. M., and Belshakov, D. A., *Neftehimiia*, **4**, 535 (1964)
363. *European Chemicals News*, **3**, 57, 25 (1963)
364. Bolshakov, D. A., *Him. Prom.*, **8**, 7 (1961)
365. Turkeltaub, I., *Zavodskaia Lab.*, **23**, 1120 (1957)
366. *Revue des Prod. Chim.*, **69**, 1351, 615 (1966)
367. Reis, T., *Chem. and Proc. Eng.*, **51**, 3, 65 (1970)
368. *Chem. Eng. News*, **41**, 47 (1963)
369. *Chem. Proc.*, **44**, 4, 40 (1962)
370. Kroper, H., Weitz, H. M., and Wagner, V., *Hydroc. Proc. Petr. Ref.*, **41**, 11, 191 (1962)
371. Ibragimov, I. A. and Mett, M. S., *Izv. Vysshikh. Uchebn. Zavedenii, Neft i Gaz*, **8** (9), 97–102 (1965) (Russ.)
372. Smith, J. and Caulier, M., French Patent, 1,485,706 (1967)
373. Esso Res. and Eng. Co., British Patent, 1,120,452 (1968)
374. May, M. and Long, R. B., U.S. Patent, 3,395,195 (1968)
375. Long, R. B., U.S. Patent, 3,412,173 (1968)
376. Gottfried, N., British Patent, 1,161,645 (1969)
377. *Chem. Week*, **86**, 22, 61 (1960)
378. Hrncirik, F., *Chem. Technik.*, **4**, 207 (1962)
379. Happel, I., Cornell, P. W., Eastman, Du. B., Fowle, Jr., M., Porter, C. A., and Schutte, A. H., *Trans. Inst. Chem. Eng.*, **42**, 189 (1946)
380. Nettesheim, G., *Erdol. u. Kohle-Erdg.-Petroch.*, **23**, 3, 156 (1970)
381. Conrad, R., *Chem.-Ing.Techn.*, **42**, 24, 1564 (1070)
382. Krönig, W., *Erdöl u. Kohle*, **16**, 520 (1963)
383. *Chem. Eng. News*, **45**, 15, 60 (1967)
384. Hart, P. M. G., *South African Chem. Proc.*, **3**, 5, 125 (1968)
385. *Chem. Eng.*, p. 78 (1969)
386. *La Chim. e L'Industria*, **45**, 665 (1963)
387. *Hydroc. Process.*, **48**, 11, 192 (1969)
388. Morrison, J., *Oil Gas Intern.*, **7**, 10, 56 (1967)
389. Farbrenfabriken Bayer, A.G., French Patent, 1,554,005 (1969)
390. Inst. of Physic. and Chem. Research, British Patent, 1,025,432 (1966)
391. Yashima, *Tatsuaki Kogyo Kagaku Zasshi*, **71**, 10, 1647 (1968)
392. *Chem. Week*, May 6, p. 73 (1961)
393. *Hydroc. Process.*, **48**, 11, 191 (1969)
394. Anhorn, V. J., French, K. J., Schafell, G. S., and Brown, D., *Chem. Eng. Progr.*, **57**, 5, 43 (1961); Garnor, J. J., U.S. Patent, 3,246,046 (1960)
395. *Chem. Week*, 29 Oct., p. 39 (1960)
396. Shell International Res. Maatschppij, N.V., British Patent, 867,296 (1961)
397. Weskow, H. Z., Colling, P. M., and Kankalits, D. C., U.S. Patent, 3,440,299 (1969)
398. Weskow, M. Z., Colling, P. M., and Karkalits, O. C., U.S. Patent, 3,428,703 (1969)
399. Adams, C. R., Preprints of the Division of Petr. Chem. Am. Chem. Soc., 14, 3, C6 (1969)
400. Voge, H. H. and Adams, C. R., *Advan. in Catalysis*, **17**, 151 (1967)
401. Adams, C. R., *J. Catalysis*, **11**, 96 (1968)

402. Pasternak, I. S. and Vadekar, M., *Can. J. Chem. Eng.*, **48**, 2, 212 and 216 (1970)
403. Timosenko, V. I. and Buianov, R. A., *Kin. i Kat.*, **12**, 1, 102 (1971)
404. Voltz, S. E. and Weller, S. W., *J. Phys. Chem.*, **59**, 569 (1955)
405. Vadekar, M. and Pasternak, I. S., U.S. Patent, 3,403,192 (1968)
406. Adams, C. R., U.S. Patent, 3,299,155 (1967)
407. Sherwood, P. W., *La Chim. e L'Ind.*, **44**, 12, 1406 (1962)
408. FIAT Final Rep., 1125
409. Sherwood, P. W., *Ind. Chemist*, 242-6 (1963)
410. McDonald, D. W., Taylor, K. M., and Brown, D. M., *Pet. Ref.* **40**, 7, 145 (1961)
411. *Pet. Ref.*, **42**, 11, 138 (1963)
412. Huntley, E. B., Kruse, J. M., and Way, J. W., U.S. Patent, 3,141,902 (1964)
413. *La Chim. e L'Ind.*, **50**, 3, 358 (1968)
414. *Informations Chimie*, **88**, 85 (1970)
415. Gianfranco Alesso, *La Chim. e L'Ind.*, **50**, 4, 457 (1968)
416. *Informations Chimie*, **88**, 93 (1970)
417. Gelkstein, A. I., Stoeva, S. S., Kulkova, N. V., and Bukshin, Y. M., *Neftchimiia*, **4**, 6, 906 (1964)
418. Gelbstein, A. I., Stoeva, S. S., Kulkova, N. V., Bukshin, Y. M., and Lapidus, V. L., *Neftchimiia*, **5**, 1, 118 (1965)
419. Bleigenberg, A. C. A. M., Lippens, B. C., and Schuit, G., *J. Catalysis*, **4**, 5, 581 (1965)
420. Erman, L., Ya, Galperin, E. L., Kalchin, I. K., Dobrzhauskii, G. F., and Chernishev, K. S., *Neorg. Him.*, *SSSR*, **IX**, 9, 2174 (1964)
421. Idol, Jr., J. D., British Patent, 867,438 (1958)
422. Callahan, J. L. and Foreman, R. W., U.S. Patent, 3,044,966 (1962)
423. Polukarova, Z. M., Shatalova, I. G., Yusupov, R. K., Kolchin, I. K., Margolis, L. Ya, and Shchukin, E. D., *Neftchimia*, **8**, 6, 899 (1968)
424. Boreskov, G. K., Van Yaminov, S. A., Dzisko, V. A., Tarasova, D. V., Dindoin, V. M., Sazonova, N. N., Olen'Kova, I. P., and Kefel, L. M., *Kin. i Kat.*, **10**, 6, 1350 (1969)
425. Shalkina, L. V., Kelchin, I. K., Margolis, L. Y., Ermalenko, N. F., Levina, S. A., and Malashevich, L. N., *Kin. i Kat.*, **12**, 1, 242 (1971); *Him. Prom.*, **1**, 21 (1971)
426. Grasselli, R. K., Smesh, D. D., and Knox, K., *J. Catalysis*, **18**, 3, 356 (1970)
427. *Europ. Chem. News*, **14**, 341, 28 (1968)
428. Schonbeck, R., *Hydroc. Process*, **48**, 8, 124 (1967)
429. Faith, W. L., Keyes, W. L., and Clark, R. L., John Wiley & Sons Inc. (1965), *Ind. Chemicals*, Ed. III, pp. 37-43
430. *Hydroc. Process.*, **46**, 11, 142 (1967)
431. *Chemische Industrie*, **2**, 94 (1969)
432. Konig, H. and Straberger, F., U.S. Patent, 3,232,977 (1963); 3, 264, 197 (1963)
433. Schonbeck, R., Konig, H., Krzemicki, K., and Kahofer, L., *Chem. Ing. Technik*, **38**, 7, 701 (1966)
434. Idol, Jr., T. D., Tiffan, A. J., and Beals, E. J., U.S. Patent, 3,005,517 (1961)
435. Borrel, M. and Newman, F. C., French Patent, 1,392,565 (1955)

436. Snam, S. P. A., French Patent, 1,367,388 (1963)
437. Bellinger, F. J., British Patent, 901,555 (1962)
438. Scherhag, B. and Hausweiler, A., French Patent, 1,368,513 (1963)
439. Gevidalli, G., Nenz, A., and Caporali, G., *La Chim. e L'Ind.*, **49**, 8, 809 (1967); *Hydroc. Process.*, **48**, 11, 147 (1969)
440. *Hydroc. Process. Pet. Ref.*, **42**, 11, 139 (1963)
441. Veatch, F., Callahan, J. L., Idol, Jr., J. D., and Milberger, E. C., *Chem. Eng. Progress*, Oct. p. 65 (1960)
442. Veatch, F., Callahan, J. L., and Milberger, E. C., *Pet. Ref.*, **41**, 11, 187 (1962)
443. *Pet. Ref.*, **38**, 7, 264 (1959)
444. *Hydroc. Process.*, **48**, 11, 146 (1969)
445. *Revue des Prof. Chim.*, **70**, 1363, 512 (1967)
446. *Chem. Econ. and Eng. Review*, **5**, 67 (1970)
447. De Graf, W. N., *S. Afr. Chem. Process*, p. 113 (1968)
448. Stueben, K., *High Polym*, **24**, 1-80 (1970)
449. Ritchie, P. D. J., *J. Chem. Soc.*, p. 1054 (1935)
450. Burns, R., Jouls, D. T., and Ritchie, P. D., *J. Chem. Soc.*, p. 714 (1935)
451. Fairweather, H. G. C., American Cyanamid Co., British Patent, 569,470 (1945)
452. Grassie, N. and Vance, E., *Trans. Faraday Soc.*, **52**, 727 (1956)
453. Gotkis, D. and Cloke, J. D., *J. Amer. Chem. Soc.*, **56**, 2711 (1934)
454. Kern, W. and Fernow, H. J., *Prakt. Chem.*, **160**, 246 (1942)
455. Kantier, C. T. and Heinz, A. R., U.S. Patent, 2,210,320 (1941)
456. Wagner, C. R., U.S. Patent, 2,412,437 (1946)
457. Heinermann, H., U.S. Patent, 2,672,477 (1945)
458. Davis, S. H. and Carpenter, E. L., U.S. Patent, 2,500,403 (1950)
459. Dalin, M. A., *Dokl. Akad. Nauk. SSSR*, **154**, 115 (English) (1964)
460. Brill, W. F. and Finley, I. H., *Ind. Eng. Chem. Prod. Res. Dev.*, **3**, 2, 89 (1964)
461. Koening, H., Schoenbeck, R., and Straberger, F., Austrian Patent, 226,672 (1963)
462. Farbenfabriken Bayer, A.G., British Patent, 957,022 (1964)
463. Hadley, D. J., Bream, J. B. and Bethell, J. R., German Patent, 1,165,015 (1964)
464. Bethell, J. R., Barclay, J. L., Madley, D. J., Stewart, D. G., and Wood, B., German Patent, 1,165,584 (1964)
465. Toyo Rayon Co. Ltd, French Patent, 1,355,561 (1964)
466. Jennings, T. J. and Voge, H. H., Belgian Patent, 618,298 (1962)
467. Bream, J. B., Hardley, D. J., Barclay, J. L., and Stewart, D. G., British Patent, 876,446 (1959).

Index